THE ENCYCLOPEDIA OF THE
EARTH

THE ENCYCLOPEDIA OF THE
EARTH

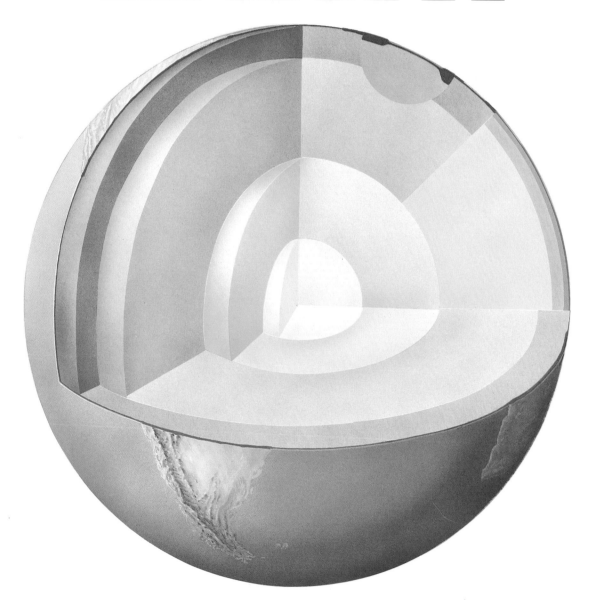

ANDROMEDA

THE ENCYCLOPEDIA OF THE EARTH

Consultant Editor: Robin Kerrod
Art Director: John Ridgeway
Designer: David West
Text Editors: Steve Luck,
    Caroline Sheldrick
Production: Steve Elliott

Media conversion and typesetting:
Peter MacDonald, Una Macnamara and
    Vanessa Hersey

Planned and produced by:
Andromeda Oxford Ltd
11–15 The Vineyard, Abingdon,
Oxfordshire OX14 3PX

Copyright © Andromeda Oxford Ltd 1993

ISBN 1 871869 11 0

Published in Great Britain by
Andromeda Oxford Ltd
This edition specially produced for
Selectabook Ltd

Origination by Alpha Reprographics Ltd,
England

Printed in Hong Kong by Dai Nippon Ltd

Authors:
Dougal Dixon
John Stidworthy
Clint Twist

Contents

Introduction

The planets of our Solar System were born from a cloud of dust and gas nearly five billion years ago. The Earth evolved quite differently from the other planets, however, making it possible for life to flourish in a variety of different forms. *The Encyclopedia of the Earth* provides a fascinating insight into the physical evolution of the Earth from its origins through to the animal and plant communities that live on it today. The encyclopedia is divided into clearly defined, self-contained sections, preceded by an introduction to each one. Each section can be read and enjoyed by itself, or can be read as part of a progressive understanding of our planet.

The Earth consists of a heavy central core surrounded by many different layers. The inner part of the core is liquid, the outer part of the core is solid. The surface, or crust, of the Earth is made up of plates which form continents and seas. Each plate is gradually moving and, over a period of millions of years, the surface of the Earth has undergone frequent and often violent changes as they have pushed against one another and then moved apart again. The first sections of this encyclopedia examine the Earth's formation and structure and reveal the sometimes devastating effects of the movement of the Earth's plates.

The crust of the Earth consists of several types of rock. They have all been formed in different ways. Their formation often determines their structure and hardness. *The Encyclopedia of the Earth* examines the different rocks that exist, and how, by studying them, we can better understand the history of our planet.

The rocks that make up Earth's landscape are constantly being altered by the erosive effects of wind and water. Using clear artwork and carefully chosen photographs, *The Encyclopedia of the Earth* traces these changing landforms to show how, for example, underground caves are created or how desert sand dunes are formed. In this way, the reader obtains a clear and comprehensive introduction to understanding today's landscape.

From space, the Earth looks blue. This is because of the effect of Earth's atmosphere and because more than two-thirds of its surface area is covered in water. *The Encyclopedia of the Earth* examines the seas and oceans; their composition, the life they contain and the effects that they have on our climate. Also explored is Earth's atmosphere; its components and structure, and how it is gradually changing.

Weather and climate are fundamental both to our lives and to our understanding of the nature of our planet. Although the differences between weather and climate are often confused, by using clear diagrams and photographs, *The Encyclopedia of the Earth* conveys these differences. It also explains weather systems, the increasingly sophisticated ways in which weather forecasting data is gathered, and the climates of the world including the rainforest climate, mountain climates and the extremes of desert and polar climate.

After describing the Earth's physical geography, *The Encyclopedia of the Earth* looks at the origins of life on Earth and the evolution of the plant and animal communities over the millennia. This traces change from the simple life forms over 500 million years ago, through the Age of Dinosaurs 213–65 million years ago, to the ascendancy of mammals and Man. Also covered are the forces of natural selection and evolution that continue to shape the various forms of life on Earth. It then profiles the present day life forms from the simplest protozoans and algae to birds and mammals. Each group has its own amazing stories to tell – such as the electric eel which can produce 400 volts to stun its prey or the tiny bodies of coral polyps which have constructed the 1,900 kilometre long Great Barrier reef – the largest animal-made structure in the world.

The balance between the various forms of life on our planet is delicate: how we exploit our planet is essential not only to our own survival but to the survival of all other life and, ultimately, to Earth itself. The last sections of *The Encyclopedia of the Earth* look at how we make use of the plants, animals and natural resources that the Earth provides. It shows how we have developed agriculture around the world. Yet, despite such technological advances, between 60,000 and 80,000 people on average still starve to death each day. The disadvantages of modern agricultural techniques and the use of chemicals are analysed: they may have increased yields for some farmers but is this approach storing up even greater problems for future generations as water

sources become polluted and exhausted and soil erosion increases?

It is widely felt that the Earth today is at a critical point in its history. The last sections of *The Encyclopedia of the Earth* provide an introduction to the problems and solutions regarding the state of the planet in relationship to man and look at issues affecting the balance of the planet's ecosystem including such topics as species extinctions, global warming, acid rain and pollution.

The Encyclopedia of the Earth has been designed to deliver vast amounts of information clearly and comprehensively. The rich assortment of images and concepts of and about the Earth will enable people of all ages to understand the geology of our planet and appreciate its beauty and fragility. By instilling a sense of wonder and of diversity, *The Encyclopedia of the Earth* will further inspire a sense of responsibility towards our planet.

Spot facts and fact boxes
Each section of *The Encyclopedia of the Earth* is prefaced by a short introduction and a series of memorable "spot facts" which highlight some of the subjects covered in the pages that follow. Scattered throughout the book are fact boxes which focus on particular subjects to help reinforce the general text – for example, on plant succession or the geography of the ocean floor.

The planet Earth

Spot facts

• *The Earth consists of different concentric shells, like the layers of an onion. The main layers are the crust, mantle and core.*

• *Only one four-hundredth of the Earth's mass consists of continental crust, which we can examine. Continental crust is on average five times thicker than oceanic crust, which is difficult to examine.*

• *The mass of the Earth is about 6 million million million million kg.*

• *The volume of the Earth is about 1,000 million million million cubic metres.*

• *The density of the Earth is about 5.5 times the density of water.*

The Earth, our home planet, is quite unlike any other in the Solar System. It was formed at the same time as the rest of the Solar System. It is a rocky planet like Mercury, Venus and Mars, the other inner planets. However, the Earth's distance from the Sun gives it the conditions that can allow water to exist and first enabled life to develop. The Earth spins on its axis and moves around the Sun. These motions give us, respectively, days and years. The tilt of its axis creates the seasons. Modern scientific techniques allow us both to investigate the Earth's interior, and to survey the surface of the planet.

▶ Steam arises above an area of geysers and hot springs. These occur in volcanic regions where the Earth's outer skin, or crust, is weak or the forces underneath are very strong.

The unique Earth

The Sun is at the centre of a whole family of planets. Close to it are the inner planets – Mercury, Venus, Earth and Mars.

Mercury, closest to the Sun, is an airless, cratered rock not much larger than our Moon. It is a blisteringly hot cinder of a world totally incapable of supporting life of any kind. If it once had an atmosphere, this would have disappeared into space long ago.

Venus is the next planet out from the Sun and is about the same size as the Earth. Whereas Mercury is airless, Venus is clothed in a thick atmosphere of mainly carbon dioxide. At the surface, the pressure of the atmosphere is nearly 100 times that of Earth. The thick blanket of gases acts like a greenhouse and traps the Sun's heat, and gives a surface temperature as high as 480°C. There can be no life on Venus.

Mars, the planet beyond us, is smaller than the Earth. Mars also has an atmosphere, but only a very low pressure one. Again, the atmosphere consists largely of carbon dioxide.

Water exists on the surface, but only in the form of ice at the poles. Pressures and temperatures are never high enough to make all the ice melt. The whole planet is a red, lifeless desert.

Beyond Mars lie the outer planets – the gas giants of Jupiter, Saturn, Uranus and Neptune. These are far larger than the Earth, and their outer visible·structure consists of the gases hydrogen and helium. They are so unlike the Earth and the other inner planets, in structure and composition, that it is difficult to make comparisons. Beyond them, the outermost planet Pluto is so distant that it is something of a mystery altogether.

Among all the planets, only the Earth is of such a size and at such a distance from the Sun that its surface is neither too cold nor too hot. Because of the temperature and pressure, water can exist in all its three forms – as gas, liquid and solid. These conditions also allow plants and animals to reproduce and evolve. In short, Earth is the only planet that has the conditions that support life.

▲ Photographs sent back from the Martian surface by the *Viking* space probes in 1976 showed a red stony desert with a red dusty sky. No sign of life was found. The low temperatures and pressures make Mars very unsuitable for living things.

◀ The Earth has blue skies filled with clouds, and in most places there is standing or running water. Life abounds, with plants using sunlight and water to make food. Plants in turn support the complex web of animal life across the globe.

Earth's origins

Since the dawn of history people have wondered about how the Earth came to exist. Many theories were put forward, based on what little knowledge was available at the time. Most of these are now out of date. Nowadays, with the use of modern technology, we are continually amassing more and more information about the Solar System and how the Earth was born. We use satellites, space probes and scientific instruments to look at the stars and the planets. We are gradually finding out what makes the stars give out light and what the planets are made of. Scientists use this knowledge to find out more about our own Earth.

The most up-to-date theory suggests that the whole Solar System – the Sun, the Earth and the rest of the planets, moons and asteroids – formed from a single mass of cold dust and gas called a nebula. About 4,600 million years ago the nebula began to shrink. This was caused by the force of gravity, which pulls objects together because of their mass.

While this process continued, the Sun began to form. As it shrank, the nebula began to spin, which made it flatten out into a disc. The spinning at the centre was greater than towards the outside of the disc, and the outermost parts sheared off as rings. The material at the centre heated up rapidly, and then began to shine as the Sun.

The planets could have formed when the gas and dust in the rings gathered together into solid bodies. Or they may have been built up from layers of dust.

Formation of the planets

There are two theories about how the Solar System may have formed. It probably started out as a nebula, consisting of clouds of dust and gas, which contracted to form the Sun and planets. According to the accretion theory (1), the force of gravity caused lumps of matter to become welded together. They grew into larger and larger bodies until they became planets. The alternative proto-planet theory (2) suggests that dust from the nebula gathered in various gravitational centres and then gathered together to form the planets.

1

2

Gravity and magnetism

A tremendous force caused the original nebula to contract and form the Solar System. This was the force of gravity. It is difficult to define, but it acts on all matter in the Universe. Its effect is to pull towards each other all things that have mass – planets, rocks or particles of dust. The gravitational force between the Earth and all of the objects on the Earth causes the objects to have weight.

Other major effects result from the Earth's magnetic field. The Earth acts like a giant magnet, tending to pull magnetic materials towards its north and south poles. This effect is used in a compass, which points northwards because the magnetic north pole is near the geographical North Pole. The origin of the magnetic field may be the Earth's core.

▲ The Aurora Borealis, or Northern Lights, occurs when the Earth's magnetic field traps charged particles from the Sun. The particles interact with molecules in the air and make them give off a glow.

Earth's magnetic field

The magnetic field of the Earth resembles that of a giant magnet in the centre of the Earth, pointing north and south. A compass (below) has a magnetized needle pivoted at its centre so that it can swing from side to side. The right-hand side of the diagram (below right) shows how a compass needle lines itself up with the magnetic field. As a result, it always points to the north. The magnetic and geographic poles are usually in slightly different places. A dip needle is a magnetic needle pivoted so that it can swing up and down. It also follows the magnetic field lines, and points straight up or straight down when over the poles.

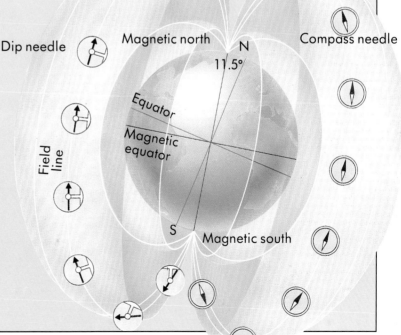

Dip needle

Magnetic north

11.5°

Compass needle

N

Equator

Magnetic equator

Field line

S

Magnetic south

The skin of the Earth

The Earth's surface is in constant motion. It is made up of a series of slabs, or "plates", which shift and drift so slowly that the movement can hardly be detected. Yet over millions of years we can see the result as continental drift – the gradual movement of the continents.

The moving part of the Earth is a stiff skin, the lithosphere, about 75 km thick. It lies on top of a softer layer called the asthenosphere. The lithosphere is like a cooling scum forming on the more plastic asthenosphere.

▶ A cross-section of the Earth shows weak points where material from the asthenosphere pushes to the surface to form new lithosphere. They occur on ridges along the ocean floor, as in the Indian Ocean Rise (1).

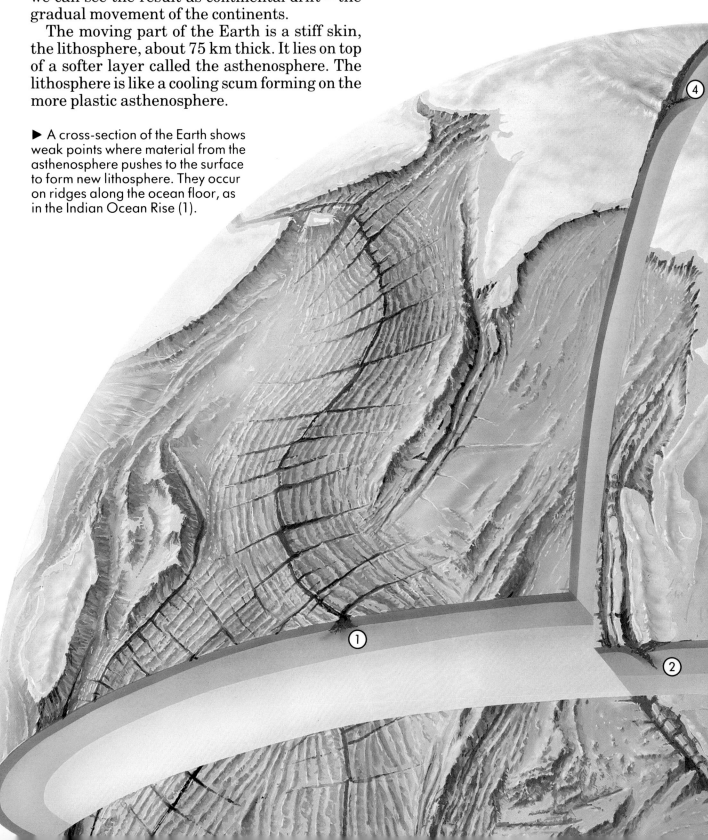

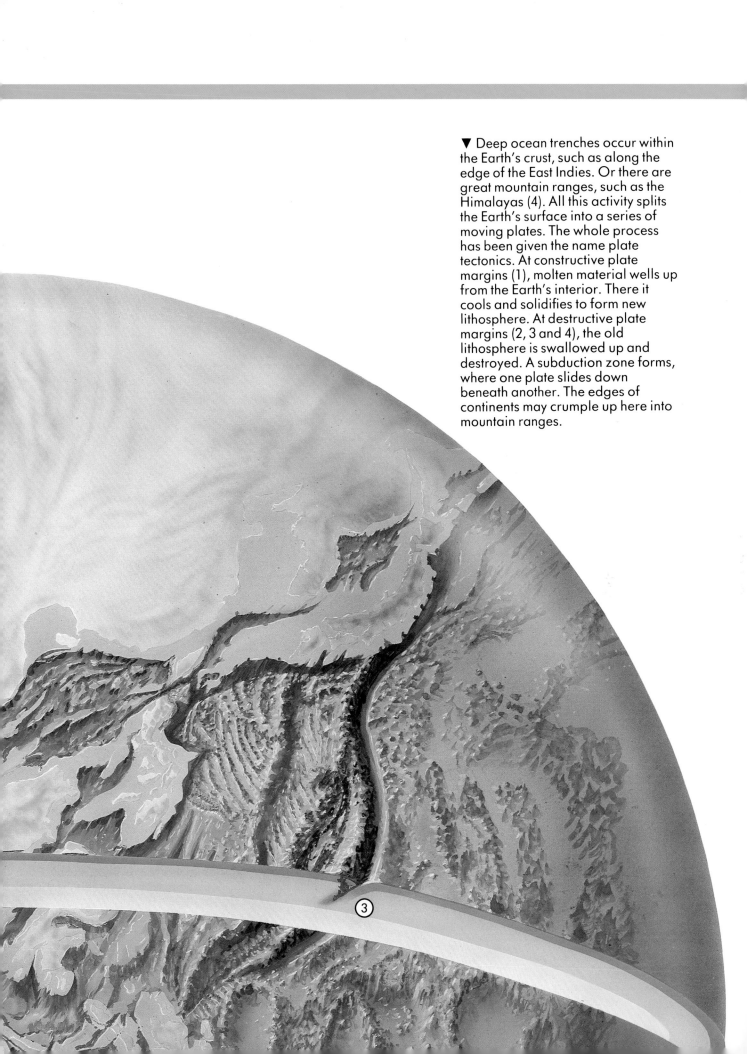

▼ Deep ocean trenches occur within the Earth's crust, such as along the edge of the East Indies. Or there are great mountain ranges, such as the Himalayas (4). All this activity splits the Earth's surface into a series of moving plates. The whole process has been given the name plate tectonics. At constructive plate margins (1), molten material wells up from the Earth's interior. There it cools and solidifies to form new lithosphere. At destructive plate margins (2, 3 and 4), the old lithosphere is swallowed up and destroyed. A subduction zone forms, where one plate slides down beneath another. The edges of continents may crumple up here into mountain ranges.

Earth's motion

At one stage when the Solar System was forming, the planets were nothing more than a series of rings of loose particles around the early Sun. It was gravitational force that held the particles in each ring. They were being pulled inwards towards the Sun, but at the same time they were moving along so fast that they were flying past it. At a certain speed the two movements balanced and the particles fell into orbit round the Sun, never reaching it but never breaking away. An orbit is a circular or elliptical path in which the tendency of a body to fly away into space is just balanced by the force of gravity pulling it inwards.

When the rings solidifed into the planets, the planets themselves were still in orbit around the Sun. Many of them had smaller bodies in orbit around them. These are known as satellites or moons.

The Earth's orbit is not circular, but is slightly elliptical. At its closest point to the Sun (perihelion) it comes to within 147,100,000 km. At its farthest point (aphelion) it is 152,100,000 km away. The Earth reaches perihelion in early January, and aphelion in July. It takes 365¼ days for the Earth to travel once round its orbit. This is the period of time we call a year.

As the Earth travels in its elliptical orbit, it also spins, or rotates, on its own axis. It does this every 24 hours, giving us our days and nights. The Earth's axis is tilted at an angle (23½°) to the plane of its orbit round the Sun. At perihelion the North Pole is tilted away from the Sun and the South Pole tilted towards it. At aphelion the tilt is reversed, the South Pole tilting away from the Sun and the North Pole tilting towards it. It is the tilt which gives rise to the change of seasons throughout the world.

Changing seasons

It is the tilt of the Earth's axis that creates the various seasons. The Earth's axis is tilted at 23½° to the plane of its orbit around the Sun. As the North Pole tilts towards the Sun in June, the Sun does not set at all in the far north, and it is northern summer. It is northern winter when the North Pole tilts away from the Sun.

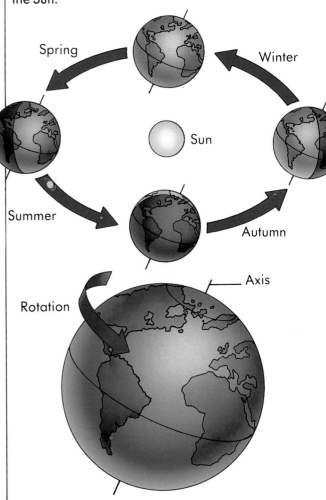

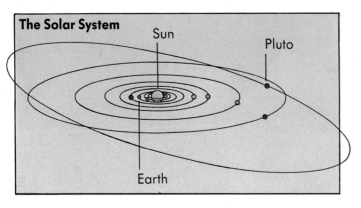

The Solar System

◄ The orbits of nearly all the planets of the Solar System lie in much the same plane. This would have been the plane of the original disc of gas that formed the Solar System. A notable exception is Pluto, which has an orbit that is tilted in relation to this plane. Pluto's orbit is also highly elliptical, and at its perihelion the planet actually lies inside Neptune's orbit.

▼ When seen from space, the Earth looks perfectly round. The deepest ocean basins and the highest mountains hardly blemish its spherical surface. But the Earth is slightly flattened at the North and South Poles. As the Earth spins on its axis, the rotation makes it bulge outwards along the Equator.

Structure of the Earth

As the Earth solidifed from its nebula of gas and dust, its components separated themselves according to their densities. Perhaps the densest substances congregated first, and then the less dense ones gathered on the outside. This would have resulted in a structure with a heavy central core and a lighter covering. Alternatively, all the substances may have clumped together at the same time. Then the densest of them would have sunk through to collect at the centre. Whatever happened, we now have a planet with a massive core covered by a number of different layers.

At the centre lies the core. It is made of the heavy metals iron and nickel. There are two layers here. The inner core is solid, whereas the outer core is liquid.

Around the core lies the rocky mantle, which also consists of two layers. The mantle makes up the bulk of the Earth. It is mostly solid and is made of silicates, compounds of silicon and oxygen. Most of the Earth's rocks are made up of silicates.

On the outside is the crust. This is, to us, the most important of the layers, and it is the only one we can reach directly. There are two types of crust, made of slightly different silicate materials. The larger area consists of oceanic crust, which is quite thin. It is made largely of silica and magnesium, and is given the shorthand name sima. The second type of crust forms the continents. It is made mostly of silica and aluminium, and is called sial.

Sial is lighter and thicker than sima. The continents are formed of separate lumps of sial "floating" in the sima of the ocean floor. Unlike the sima, the sial is not carried downwards. As a result, the continents are much older than the ocean floors.

The layers of lithosphere and asthenosphere which make up and move the plates of the Earth's surface, are found towards the outside of the globe. The solid lithosphere consists of the crust and the uppermost part of the mantle. The soft asthenosphere is a distinct layer of mantle positioned just below it.

▼ Pillow lavas are found on the seabed. They are cushion-shaped lumps of volcanic material formed when lava erupts from a submerged ocean ridge and cools quickly. The whole of the deep ocean floor, beneath the sediments, is formed of pillow lavas. They are part of the sima – the material of the oceanic crust. Sima is rarely seen because it lies beneath several kilometres of ocean water and sediments.

▼ All the minerals and rocks of mountains and plains are components of the sial. The sial – the material of the continental crust – is all around us. It tends to be more complex than the sima, because it is much older. Its rocks are constantly being lifted up by mountain-building activity and worn away by weather and rivers. The fragments are redeposited as sediments, and eventually they are turned into solid rocks once more.

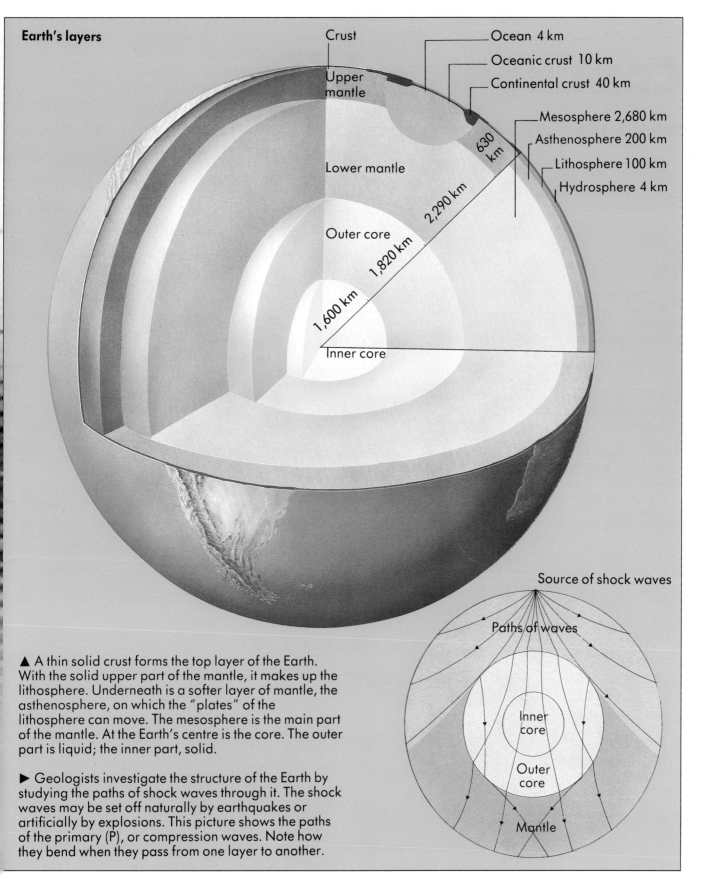

Earth's layers

Crust

Upper mantle

Ocean 4 km

Oceanic crust 10 km

Continental crust 40 km

Mesosphere 2,680 km

Asthenosphere 200 km

Lithosphere 100 km

Hydrosphere 4 km

630 km

Lower mantle

2,290 km

Outer core

1,820 km

1,600 km

Inner core

Source of shock waves

Paths of waves

Inner core

Outer core

Mantle

▲ A thin solid crust forms the top layer of the Earth. With the solid upper part of the mantle, it makes up the lithosphere. Underneath is a softer layer of mantle, the asthenosphere, on which the "plates" of the lithosphere can move. The mesosphere is the main part of the mantle. At the Earth's centre is the core. The outer part is liquid; the inner part, solid.

▶ Geologists investigate the structure of the Earth by studying the paths of shock waves through it. The shock waves may be set off naturally by earthquakes or artificially by explosions. This picture shows the paths of the primary (P), or compression waves. Note how they bend when they pass from one layer to another.

19

Continents adrift

Spot facts

• *The longest mountain range is the Andes in South America – 8,900 km long.*

• *The highest mountain is Mount Everest at 8,848 m.*

• *The biggest continent is Europe and Asia combined – Eurasia – with a land area of 54,750,000 square kilometres.*

• *50 million years ago Australia was joined with Antarctica.*

• *The Atlantic Ocean is 10 m wider today than it was when Christopher Columbus discovered America.*

▶ The Himalayas are the highest mountain range on Earth. They were formed quite recently – within the last 50 million years – by the northward movement of the plate that carries India.

The surface of the Earth is an unstable, shifting place. Since the mid-1960s it has been known that the crust and the topmost layer of the mantle form a number of distinct plates that cover the Earth like the panels of a football. The plates are being generated continuously along one edge and destroyed along another. The shifting of continents across the globe, something suspected for centuries, is the result of all this movement. Earthquakes and volcanic eruptions are two destructive side-effects that occur at plate boundaries.

Dynamic Earth

The greatest mountain ranges on Earth – the Himalayas, Rockies, Andes, Alps and Urals – were formed by a folding action of the upper layers of the Earth's crust. As a result, the rocks are twisted and bent, obviously deformed under some great pressure. It is the movement of the Earth's surface plates – the action of plate tectonics – that has brought this about.

Each plate is created from molten material from the Earth's interior. This wells up in volcanic activity along the great ridges that run along the ocean beds. The plates build out from their edges and move apart. Eventually, when a plate meets another plate travelling in the opposite direction, one plate slides beneath the other and is destroyed.

Sometimes this destructive plate margin lies along the edge of a continent, as it does along the western coast of South America. Then the edge of the continent is crumpled up into a mountain range – in this instance the Andes. Often two continents collide and produce a range where the two continents have fused. The Himalayas and Urals formed in this way.

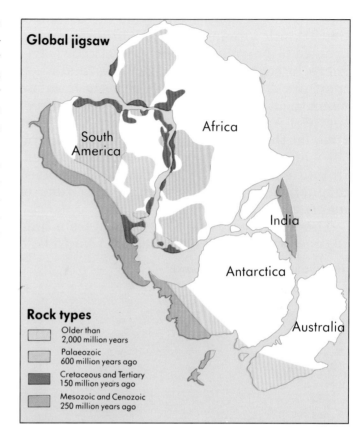

Global jigsaw

South America

Africa

India

Antarctica

Australia

Rock types

- Older than 2,000 million years
- Palaeozoic 600 million years ago
- Cretaceous and Tertiary 150 million years ago
- Mesozoic and Cenozoic 250 million years ago

▲ Continents may split apart as new constructive plate margins develop beneath them. South America, Africa, India, Antarctica and Australia were once a single landmass. They broke up and drifted apart as new oceanic crust developed between them. We know this because of the shapes of the continental shelves, the similarity of fossils on each landmass, the continuation of mountain ranges across them and similar rock types.

◄ The intensely folded rocks of Wales are the result of past movements. Once there was another ocean where the Atlantic now lies. This closed up as the landmasses on the east and west collided. They threw up a mountain range like the Himalayas. After many millions of years the continents broke apart again and the modern Atlantic Ocean formed between. The remains of the old mountain range lie in the Appalachians in North America, and the Welsh, Scottish and Norwegian highlands in Europe.

Surface plates

In the 1960s, scientists studying the bottom of the Atlantic Ocean began to notice something that led to a revolution in geological knowledge. All rocks contain magnetic particles, and when they are formed, their magnetism lines up along the Earth's magnetic field. At the crest of the Mid-Atlantic Ridge, the ocean ridge that runs north to south along the Atlantic Ocean, the magnetism of the rocks lines up as expected. But to each side the rocks have a reversed magnetism, evidently produced at a time of magnetic reversal.

These magnetic reversals happen from time to time, when the Earth's magnetic North Pole and South Pole change polarity. Further exploration showed that down the sides of the ridge the rocks showed alternating stripes of normal and reversed magnetism. The pattern on one side was the mirror image of the pattern on the other. This indicates that the seabed is being created along the crest of the ridge, and is moving away to each side as new material continues to well up from below. As each band of new rock is formed, it lines up with the direction of the Earth's magnetic field at the time. This process is called seafloor spreading.

Further proof that the rocks of the seabed are spreading came when it was noticed that the rocks on the ridge of the crest are fresh, whereas those farther away are covered in sediment. The sediment layer becomes thicker even farther from the crest, reflecting how much older the seabed must be.

Soon this new concept of seafloor spreading was put together with the older discovery of continental drift to produce the new science called plate tectonics.

The changing face of the Earth

150 million years ago

100 million years ago

50 million years ago

▲ The most obvious result of plate tectonics is that of continental drift. Throughout geological time the continents have been carried across the face of the Earth, embedded in the plates like logs locked in the ice of a frozen river. There is enough evidence to plot the position of the continents at different stages of the Earth's history.

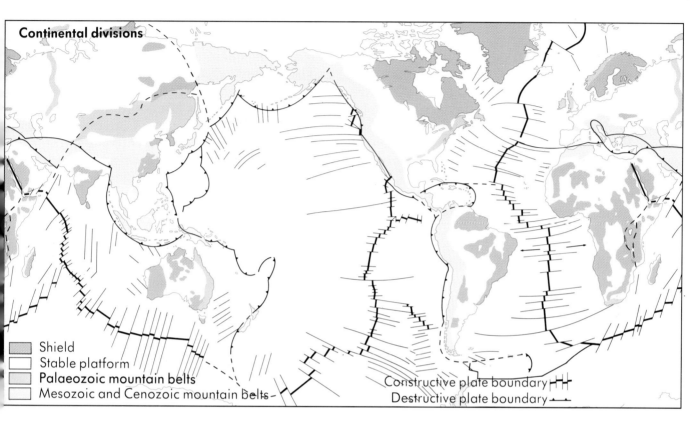

Continental divisions

- Shield
- Stable platform
- **Palaeozoic mountain belts**
- Mesozoic and Cenozoic mountain belts

Constructive plate boundary ┤┤┤
Destructive plate boundary ▲━━

▲ Continents are made up of three basic components: the shield, a flat plain of rock; the stable platforms, the base of which is made up of the same kind of rock as the shield, but with a layer of different rock on top; and the mountain belts. The oldest mountains – the Palaeozoic – are nearest the shield, and the younger – the Mesozoic and Cenozoic – are farthest away from it.

Today

Wegener the pioneer

In 1912 the German meteorologist Alfred Wegener (1880-1930) came up with the idea of continental drift based on sound scientific reasoning. He produced a series of maps of the world as it was in the past. They were essentially the same as those that can be produced today with our vastly increased knowledge. However, he could not account for the mechanics of the movement, and did not live to see plate tectonics provide the answer.

Rifts and mountains

All the highlands and lowlands, mountain ranges and plains, and the whole large-scale landscape of the world today can be looked at as the result of the activity of plate tectonics.

The most extensive mountain ranges are those of the fold mountains. They are caused by compression – by two plates grinding into each other. When two plates, topped with oceanic crust, meet each other, one is destroyed. It slides down into the depths of the mantle, pulling down the seabed into an ocean trench, while the other plate rides up above it. As the descending plate melts, the molten material rises up through the overlying plate and bursts through to the surface as a series of volcanoes. These grow until they rise above the surface of the sea as an arc of volcanic islands. Many such island arcs, running parallel to ocean trenches,

are found along the northern and western fringes of the Pacific Ocean.

A slab of continental crust may be embedded in the oceanic crust of the plate that is being destroyed. Eventually this continent reaches the ocean trench at the subduction zone, and there the movement stops. Continental crust is too light to be drawn down into the mantle.

If the plate movement continues, the oceanic crust of the opposite plate then begins to slide down beneath the continent. This establishes another subduction zone and ocean trench. The edge of the continent deforms and crumples up with the activity, while sediments from the descending plate are scraped off against it. Molten material from below thrusts up through the continental edge and a series of volcanoes develops in the deformed margin. As a result, a

▶ When a continent begins to split apart, the action starts with the upwelling of material from the Earth's mantle (1). The continent above heaves upwards and splits into blocks. These may stand up as block mountains or subside as rift valleys. Molten material forces its way through the cracks (2). The sea floods in and the rift valley becomes a young ocean, with a floor of newly formed oceanic crust (3).

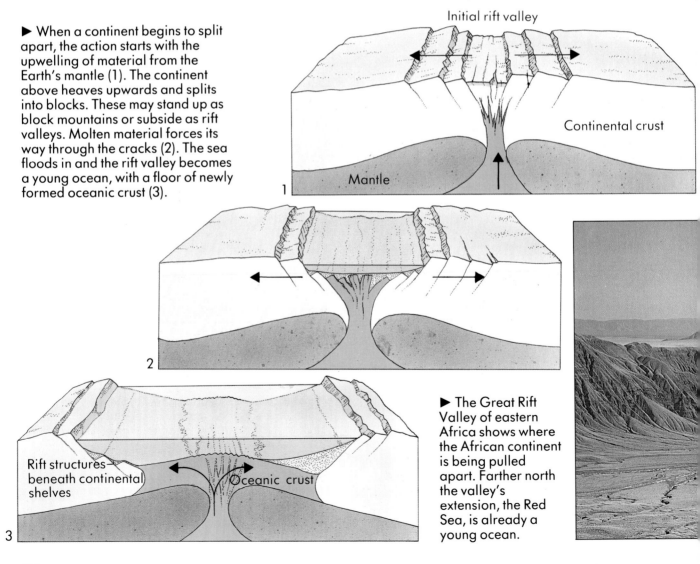

Initial rift valley

Continental crust

Mantle

1

2

3

Rift structures beneath continental shelves

Oceanic crust

▶ The Great Rift Valley of eastern Africa shows where the African continent is being pulled apart. Farther north the valley's extension, the Red Sea, is already a young ocean.

very complex chain of fold mountains grows along the continental edge. The Andes, along the western edge of South America, are a particularly good example, with an ocean trench just offshore There are also a great number of volcanoes along their length, which is another typical feature of fold mountains.

The subducting plate may bring its own continent, and when the two collide the resulting mountain range is particularly enormous, such as the Himalayas between India and Asia. Now the movement really does stop. But the forces are still at work, and make themselves felt at some other part of the globe. A new constructive plate margin develops somewhere. If it is in the middle of a continent, the continent heaves up and cracks to produce the other major type of mountain range, the so-called block mountains.

As a result all the continents have the same general pattern. There is a central area of hard old rock, usually worn flat with age, surrounded by successively younger ranges of fold mountains. A rift valley among block mountains may show where the continent is being pulled apart. If one border of the continent shows the cracks and block mountains that we would associate with a rift valley, then the continent has probably broken away from another one some time in the past.

Coastal volcanoes

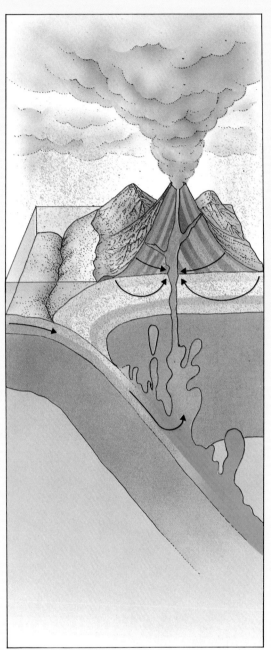

When one plate slides below the next, it is eventually destroyed in the mantle. Friction between the two plates melts the rock along the boundary, and this rises through the plate above, eventually forming volcanoes on the surface. The melting of the rocks is helped by the presence of seawater brought down by the moving plate, and a lot of water is erupted from the volcanoes.

Volcanoes

Volcanoes occur where hot material leaks out from the Earth's interior. This usually happens at the margins of the tectonic plates.

At constructive plate margins, along the ocean ridges, the hot material from the mantle wells up and solidifies. The molten matter that is forced out at the surface is called lava. It solidifies not far from the vent, gradually building up into a mountain. The activity takes place on the ocean floor, and so these types of volcanoes are rarely seen. Only in places such as Iceland does the ocean ridge reach above the surface of the water. Then the volcanoes can be seen on land.

At destructive plate margins the molten material comes from the breakdown of the plates themselves, and the volcanoes form in island arcs or in fold mountains. The lava is a different type from constructive margin lava, and forms different types of volcanoes.

A third type of volcano is found away from the plate margin, over a "hot-spot" of activity deep in the mantle. The lava that erupts is of the same type as that found at a constructive plate margin, and the same kind of volcanoes are thereby produced.

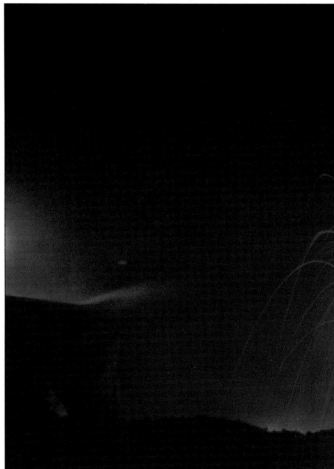

Volcanoes and earthquakes

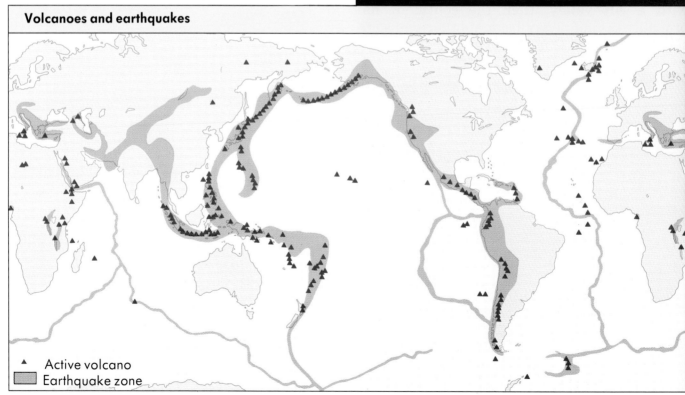

▲ Active volcano
▨ Earthquake zone

Types of volcanoes

At destructive plate margins the lava is rich in silica. This makes it stiff, producing steep-sided volcanoes such as the composite cones and cinder cones. Eruptions can be very violent, and the rock formed is called andesite. At constructive plate margins the lava contains less silica. It is runny and erupts quietly, producing broad shield volcanoes and fissures. The rocks that form are called basalts.

Composite volcano

Cinder cone

Shield volcano

Fissure

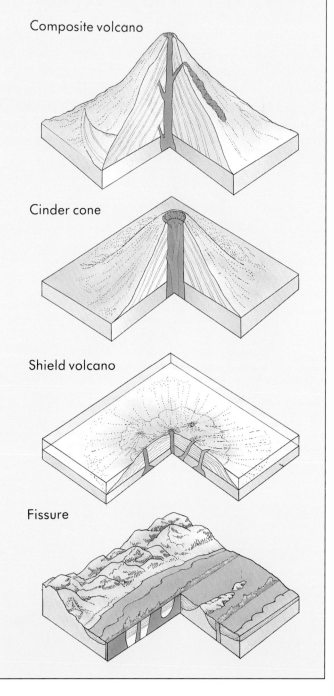

▲ A river of lava pours out from an erupting volcano. It is formed from magma from the mantle. As the magma rises and cools, some of its minerals solidify and sink back. Gases are given off as bubbles. The resulting lava does not have the same composition as the mantle.

◄ Most of the world's volcanoes are found along the margins of the plates. This distribution, as well as the distribution of earthquakes, shows where most of the plate tectonic activity is taking place.

Earthquakes

Through the action of plate tectonics the crust of our planet is always moving. Over millions of years the continents drift from one place to another. They jostle together and push up mountain ranges. The movements do not take place continuously, but in small jerks and jumps. It is these jumps that set up the vibrations we call earthquakes.

The forces that sometimes move the crust of the Earth work all the time, and cause stresses to be built up in the rocks. Eventually the stresses become so strong that they make the rocks snap. They whip along a crack, called a fault, and this movement causes an earthquake shock. The shock waves travel outwards from the focus, the point where most of the movement takes place, like the ripples from a splash in a pond. The point on the Earth's surface directly above the focus is called the epicentre. Most damage is usually done there.

When the pieces of the Earth's crust snap along a fault, they usually move too far. Later they may spring back some distance and produce aftershocks, and this may continue until the rock masses have settled. Then the stresses begin to build up once more until they are released by the next earthquake.

The different types of shock wave produced by an earthquake travel at different speeds. Observatories around the world can detect them with instruments called seismographs. By timing when waves arrive, scientists can tell how far away the earthquake was. A world network of observatories can now pinpoint any earthquake's focus.

The places most likely to have earthquakes are the edges of the Earth's surface plates, where the plates are jostling each other and being created or destroyed. But earthquakes cannot be predicted.

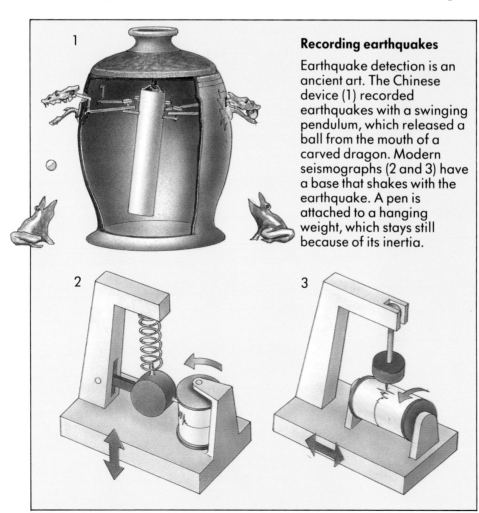

Recording earthquakes

Earthquake detection is an ancient art. The Chinese device (1) recorded earthquakes with a swinging pendulum, which released a ball from the mouth of a carved dragon. Modern seismographs (2 and 3) have a base that shakes with the earthquake. A pen is attached to a hanging weight, which stays still because of its inertia.

▶ The result of a severe earthquake that struck Mexico City in September 1985. A multistorey building has collapsed like a pack of cards. During this earthquake, more than 10,000 people died.

▼ The notorious San Andreas fault near San Bernardino in California. Movements along the fault cause frequent earthquakes.

Rocks and minerals

Spot facts

• *The most common elements in the Earth's crust are oxygen and silicon, usually united as silica. Together they make up just over three-quarters of the crust.*

• *The oldest rocks are 3,800 million years old. They are found in Greenland.*

• *Chalk is the microscopic remains of tiny creatures that lived in the seas during the Cretaceous Period of Earth's history, 144-65 million years ago. "Cretaceous" means chalky.*

• *The rocks at the top of Mount Everest formed at the bottom of the sea 50 million years ago.*

Several kinds of rocks make up the Earth's crust. They have formed in different ways: some from red-hot lava that spewed out of volcanoes, some from rock debris swept down from the mountains by rushing water and some from the fossils of ancient sea creatures. All the rocks are made up of collections of minerals, usually in the form of glassy crystals packed haphazardly together. But here and there, in cavities in the rocks, the crystals have room to grow into beautiful shapes. Well are they called "the flowers of the mineral kingdom".

► The Giant's Causeway in Northern Ireland is a basaltic lava flow 50 million years old. As it cooled and solidified, it split into the distinctive columns that make the site so famous.

Reading the rocks

The Earth's crust consists of three types of rocks. When molten material from the Earth's interior solidifies, it forms igneous rock. When fragments of sand, silt or rubble are laid down, compressed and cemented into a solid mass, the result is a sedimentary rock. When pre-existing rocks are crushed and "cooked" deep within a mountain range their mineral content changes. The resulting rocks are termed metamorphic rocks.

Sedimentary rocks are most common at the Earth's surface, and by studying them we can find out much about what happened in the past. Sediments accumulate under specific conditions. So when we see a bed of sandstone with ripple marks in it, we can deduce that it was formed from a bed of sand laid down in a shallow sea. If, above it, there is a bed of mudstone containing the fossils of freshwater snails, we can deduce that a river later swept its muds into the area and covered the sand.

Examples like this have enabled us to write the full history of the Earth's surface from the days when sediments first began to form.

◄ Two centuries of studying the rocks has enabled geologists to work out a timescale of the Earth's history. Geological time is divided for convenience into a number of eras and periods.

▼ Fossils, such as these ammonites, can help to tell the age and the history of a rock, because the animals lived only during a particular period and under certain climatic conditions.

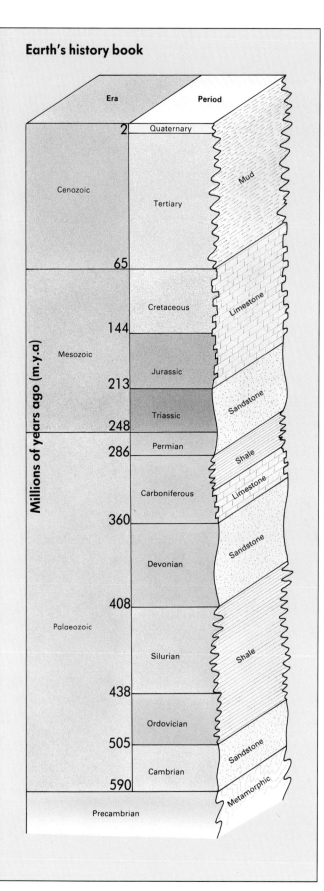

Earth's history book

Era	Period	

Millions of years ago (m.y.a)

- 2 — Quaternary (Mud)
- Cenozoic — Tertiary
- 65
- Cretaceous (Limestone)
- 144
- Mesozoic — Jurassic (Sandstone)
- 213
- Triassic
- 248
- 286 — Permian
- Shale
- Carboniferous (Limestone)
- 360
- Devonian (Sandstone)
- 408
- Palaeozoic — Silurian (Shale)
- 438
- Ordovician
- 505 — (Sandstone)
- Cambrian
- 590
- Precambrian (Metamorphic)

Changing rocks

We just need to look around us at the conditions of the Earth's surface today to realize the range of conditions that must have existed in times past. We know that we get different kinds of sands in rivers, beaches and deserts. We see that streams produce silt, and lakes become choked with peat. Coarse shingle gathers on the shoreline. At the bottom of the sea muds and oozes, and fragments of seashell and coral accumulate. All these sediments eventually produce different sedimentary rocks.

Within these rocks there are different structures. Sand laid down in a river forms twisted beds that reflect the effects of the water current. The shapes of sand dunes can be seen in sandstones formed from desert sands. Coarse sediments form from particles deposited in strong currents, and fine sediments in gentle currents. Cracks develop in drying mud, and these can be preserved in the subsequent rocks. The study of this kind of evidence is the science of historical geology, also called stratigraphy.

The rock cycle

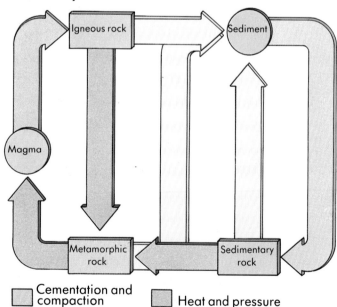

Cementation and compaction

Weathering and transport

Heat and pressure

Cooling and solidification

▲ Rocks are constantly being destroyed and recreated, a process known as the rock cycle. Many stages are involved. Sediments form sedimentary rock, which may later be crushed and re-formed in the heart of a mountain to produce metamorphic rock. The sedimentary rock may even melt with the heat and later solidify into igneous rock. All types may crumble when exposed, and their debris forms new sediments.

Movement and deposition of sediments

"The present is the key to the past," said pioneer geologist James Hutton in 1785. Using this idea, we can analyse sedimentary rocks and compare them with the many processes that are producing sediments today.

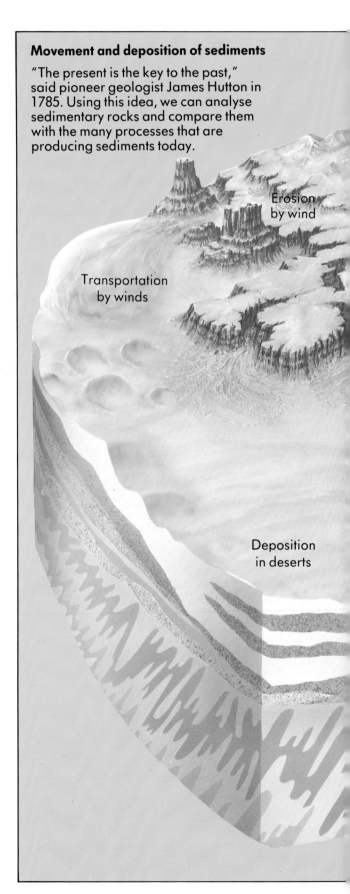

Erosion by wind

Transportation by winds

Deposition in deserts

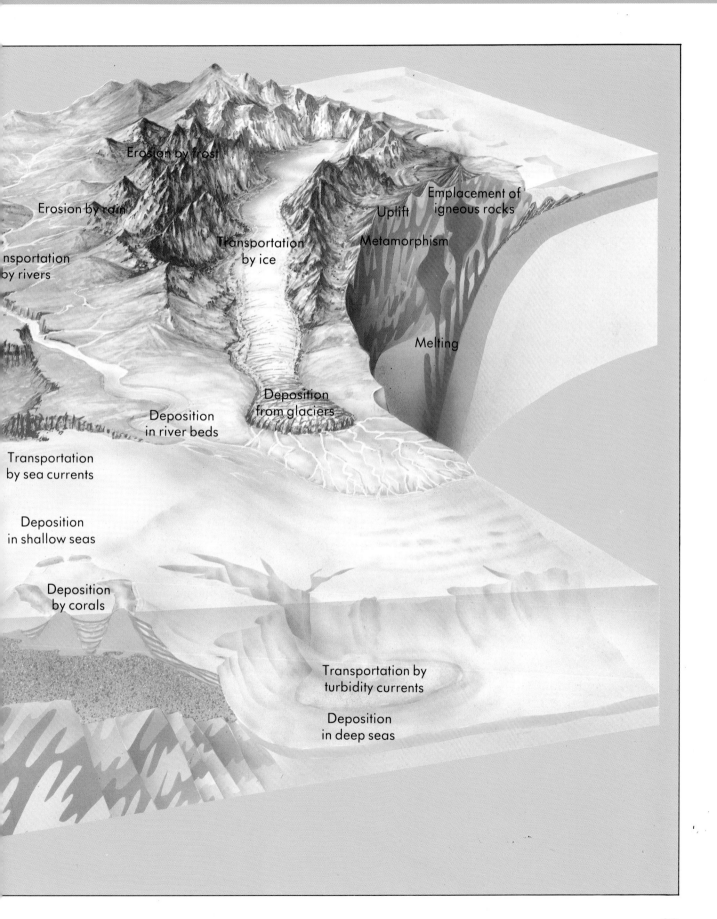

Erosion by frost

Erosion by rain

Transportation by rivers

Transportation by ice

Uplift

Emplacement of igneous rocks

Metamorphism

Melting

Deposition from glaciers

Deposition in river beds

Transportation by sea currents

Deposition in shallow seas

Deposition by corals

Transportation by turbidity currents

Deposition in deep seas

Igneous rocks

Perhaps the simplest type of rock, and the most easily understood, is igneous rock. Its formation is theoretically quite straightforward. Molten material from inside the Earth cools and becomes a solid mass.

There are two main types of igneous rock – intrusive and extrusive. Intrusive rocks form when a mass of molten material is injected into the rocks of the crust and solidifies there without reaching the surface. We see intrusive igneous rocks only when the rocks above have been eroded away. Extrusive rocks form when the molten material cools on the surface, as for instance in a lava flow.

Intrusive rocks cool very slowly, and so they tend to be coarse-grained. The crystals of individual minerals are big enough to be seen by the naked eye. Granite is a good example. Extrusive rocks cool quickly and so are fine-grained. They have microscopic mineral crystals. Basalt is an example. Sometimes the molten rock begins to cool underground and the first minerals form large crystals. Then the whole lot bursts out at the surface and solidifies quickly. The result is a rock called a porphyry, which consists of a fine groundmass with big crystals embedded in it.

Composition of igneous rocks

Igneous rocks are also classified by their composition. This is rarely the same as the original magma, the molten material from the Earth's interior. The magma is rich in the chemical silica and forms silicate minerals.

As the molten material rises through the Earth's crust and begins to cool, some minerals crystallize out before the others. The silicate minerals olivine and pyroxine are early crystallizers. These, which are rich in iron and magnesium, sink to the bottom of an intrusion. Silicate minerals such as feldspars and micas, which are low in iron and magnesium but rich in the lighter metals potassium and sodium, tend to solidify later. Uncombined silica forms the

◄ ▼ The wrinkled, ropy surface is typical of basalt, a most common extrusive basic igneous rock. After lava erupts from a volcano, it may flow for some distance over the ground as a river of fire. During this time its surface cools and hardens. The chilled surface is dragged along by the movement of the liquid beneath, and it twists and distorts as it goes. The ropy lava is known by its Hawaiian name of aa. The islands of Iceland and Hawaii were built up from the seabed by successive layers of lava flows like these.

mineral quartz. As a result, dark iron-rich rocks form deep down, whereas light-coloured silica-rich rocks form closer to the surface. Geologists call the darker rocks basic and the lighter rocks acidic.

Igneous rocks that form from magma being brought up at a constructive plate margin tend to be basic. Coarse intrusive dolerite and fine extrusive basalt are formed there. At a destructive plate margin, the magma is made of molten plate material and is usually richer in silica. The rocks that form there tend to be acidic igneous rocks, such as intrusive granite and extrusive andesite.

Although the silicate minerals are rich in metals, it is difficult to remove the metals from the silica. They are therefore no good as ores. Silicate minerals are thus referred to as the rock-forming minerals. Ore minerals – used as sources of metals – are usually sulphides or oxides of metals, and they do not usually form the bulk of the rock.

► Sheer granite cliff-faces provide a spectacular sight. El Capitan cliff in Yosemite National Park shows little sign so far of any erosion.

▼ Granite is a typical intrusive igneous rock. Its feldspar minerals break down easily on exposure to air, and so exposed granite wears away into rounded lumps.

Sedimentary rocks

The rocks which are formed when a layer of mud, sand or other natural debris is compressed and cemented together are called sedimentary rocks. Like igneous rocks, they can be classified according to their origin. The main type is clastic sedimentary rock. This is formed from fragments of other rocks, such as sand or shingle.

Any rock that is exposed at the surface of the Earth is worn away by the relentless onslaught of the wind and rain. Some of its minerals, such as feldspar, may be dissolved away by acid in rainwater. Or the rock may be broken apart by the expansion of ice in its cracks. All these actions make rock break down into fragments that can be washed away by streams or even blown away by the wind. Eventually the fragments settle. The coarsest fragments do not travel far, but come to rest as boulders, cobbles and shingle at the foot of a cliff or on a shoreline. Finer pieces like sand and silt can be carried farther and deposited on beaches or seabeds. The finest matter is washed well out to sea and settles as mud. These various sediments may eventually become sedimentary rocks such as conglomerate, sandstone and shale.

The second classification is biogenic sedimentary rock. It is formed from fragments of once-living matter, such as corals or seashells. These form limestones in which you can often see the fossils of the creatures that originally formed them.

▲ Cheddar Gorge, England, is formed of limestone. This is a very common sedimentary rock that occurs in distinct layers or beds. They may either be very thin or very thick.

▼ The sand in the foreground may eventually become sandstone like that in the background. This will only happen once the sands are buried, compressed and cemented together. Then the sandstone will appear at the surface only if the whole area is caught up in mountain-building activity, and the overlying beds are worn away. The beds that were once horizontal will be tilted up and twisted. They may be distorted by bends called folds, or they may shift along cracks called faults.

Finally, chemical sedimentary rock forms when substances dissolved in the water come out of solution and form a crust on the bottom of a lake or the sea. Rock salt and certain kinds of limestones form in this way when lakes and shallow seas dry up.

Natural concrete

When sediments are buried, they are compressed beneath the weight of the sediments on top of them. Then groundwater seeps through them, depositing mineral crystals on and between the fragments. This cements them together, just like cement holds together the gravel and sand in concrete, and turns the loose material into a hard, solid mass.

▲ Some sedimentary rocks have practical uses. The coal being mined here is a biogenic sedimentary rock made from ancient vegetation. Many sandstones and limestones make good building materials.

Sedimentary rock types

We can usually identify a sedimentary rock type by looking at the fragments that make it up. A chemical sedimentary rock, such as rock salt (1), is made up of crystals, rather like those in an igneous rock. A clastic sedimentary rock, such as the conglomerate (2), consists of distinct lumps. The lumps may be rounded if they have been washed about for a long time, or jagged and angular if they have not travelled far. They may also be coarse or fine. In a biogenic sedimentary rock, such as chalk (3), we can see the fragments of shells. The example shown here is of a microscopic fossil, but often the shell fragments can be seen with the naked eye.

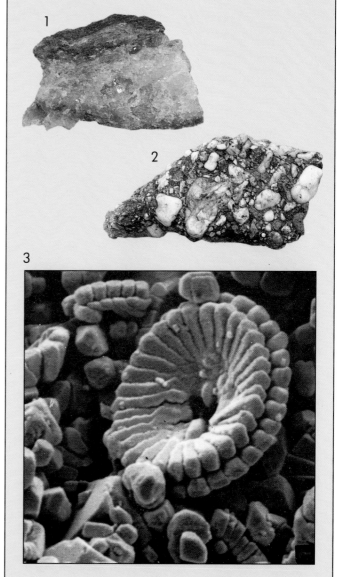

Metamorphic rocks

When the drifting continents grind into each other, the immense pressures involved crumple up mountains and alter the rocks deep inside the crust. When a rock is subjected to so much pressure and heat that its minerals change, the rock becomes a metamorphic rock. But the rock remains solid throughout this process. If it melts and then solidifies again, it changes into an igneous rock.

Geologists recognize two main types of metamorphic rock. The first is thermal metamorphic rock. This is formed principally by heat. It is usually found in localized patches around igneous intrusions, where the heat of the intrusion has re-formed the surrounding rock. A thermal metamorphic rock can be difficult to distinguish from an igneous rock, because it tends to consist of masses of inter-grown crystals with little recognizable pattern.

Thermal metamorphic rocks, however, contain minerals found in no other type of rock.

The second type of metamorphic rock is formed by pressure. The roots of whole mountain chains can be altered in this way. For this reason, the rock is called regional metamorphic rock. Slate is an example. The new minerals that form may do so in contorted layers and bands, corresponding to the direction of the pressure. The very ancient terrains in the centres of continents are usually formed from regional metamorphic rocks.

▼ A slate quarry. Slate is a fine-grained regional metamorphic rock, and one of the few that is economically valuable. The pressure that formed it produced new crystals of mica, all aligned in the same direction. Mica crystals form sheets, and for this reason slate easily splits into thin slices that can be used for covering roofs and for other purposes.

Colourful crystals

Individual minerals can be difficult to identify just by looking at them with the naked eye. Colour is not a good guide, because any mineral can contain traces of an impurity that changes the colour completely. Quartz, for instance, can be transparent, milky white, pink or brown. Corundum, an aluminium oxide, can be discoloured red, which would make it a ruby; or blue, which would make it a sapphire.

However, the "streak" of a mineral is quite distinctive. If a sample of mineral is scraped over a hard surface, it leaves a streak of fine powder. The colour of that powder is constant for a particular mineral whatever impurity it may contain.

Crystal form is a good indicator, but usually in a rock the crystals are crammed together and show no good shape.

A useful test is hardness. Some minerals are harder than others and can be tested by scratching a sample against others of known hardness. A mineral scratches only a mineral that is softer than it is. Quartz is quite a hard mineral, and it scratches softer minerals such as calcite, but can itself be scratched only by even harder minerals such as corundum.

Lustre and the effect of fracture can both be seen. When light catches the mineral, it may have a glassy, metallic or dull lustre. When it is broken, the broken face may be straight, ragged or shell-like. The appearance helps to identify the mineral.

A geologist uses a special microscope to examine a thin slice of rock and identify its minerals. The rock slice is ground until it is paper-thin and transparent. The specimen is examined using polarized light, which produces distinctive coloured patterns when it passes through a mineral.

Key
1 Pyrites, with a metallic lustre.
2 Flint, with a conchoidal, or shell-like, fracture.
3 Rock salt, with good crystal shape.
4 Quartz, with a brown impurity.
5 Diamond, the hardest mineral of all.

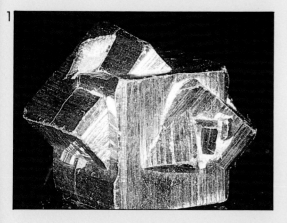

Minerals may be identified by their lustre, by the way they fracture, by the shapes of their crystals, or by their hardness.

The changing landscape

• *The highest mountains of the world are the youngest. Older ones have been worn away.*

• *Sea waves wear away the rocks in cliffs. During a storm, waves exert a pressure of up to 30 tonnes per square metre.*

• *Erosion of the Grand Canyon began in Miocene times 26 million years ago. Today it is up to 2,000 metres deep.*

• *The Amazon River carries water and sediment away from an area of more than 7 million square kilometres.*

▶ In Monument Valley, Utah, there are wide extents of thick, horizontally-bedded sandstones that have been attacked and worn away. They have been eroded by wind and rain. Everywhere we look, we can see similar changes that have taken place in the landscape.

"As old as the hills" is an expression that we use when we think of something as being very old indeed. Yet compared with the age of the Earth the hills may not be so very ancient. The Himalayas, the greatest mountain chain on Earth, are less than 50 million years old, and that is not a great span of time in geological terms. Whenever a rock becomes exposed at the surface of the Earth, and whenever an area of land rises above sea level, natural forces begin to destroy it. Gravity, running water, wind, rain, sea and frost all work together to erode, or wear down, the landscape back to sea level. The face of the Earth is also changing because of human activity.

River erosion

Much of the carving up of the landscape is done by water, and particularly by rivers. Rainwater that soaks into the ground often returns to the surface as a spring. From the spring it runs downhill as a stream.

In this young stage of a river, the water runs quickly. The faster a river flows, the more erosive power it has. It tends to erode away its bed, cutting a deep V-shaped gully as it goes. Rocks and stones picked up by the moving water are bounced along the stream bed, adding to the erosive force. Waterfalls and rapids are common at this stage.

When it leaves the hilly areas and begins to flow down a more gentle slope, the river reaches what is known as its mature stage. It still erodes the landscape but it also deposits some of the material that it has been carrying downwards. River valleys tend to be broad at this stage, much broader than the river itself. Over the years, the river's course moves about the valley floor. As a river flows round a curve, it moves more quickly on the outside. The bank at this side is undercut and worn back. On the inside of the curve the current is slower and the debris that has been carried tends to be deposited there as a beach.

The final stage of the river can be thought of as its old age. There is no valley and the water has no power to erode. It moves slowly across a plain, depositing material as it goes. Eventually it reaches the sea.

▲ A typical feature of a mountain stream is white, foaming water. It is the youthful stage of a river. It is the time when the river runs fastest and is at its most violent. It bounces over rapids and waterfalls, scouring out its narrow bed. The fast-flowing river picks up rocks and gravel and carries them along towards the lowlands. By the time it reaches the sea it is moving slowly.

Lazy river

Meanders are loops in a slow-flowing river (1). They form during a river's old age. Curves tend to get bigger by eroding the outer bank and building up the inner one (2). The result is a river that swings across a flat plain of deposited sediment, like England's River Cuckmere (3). A meander may be cut off to form an ox-bow lake.

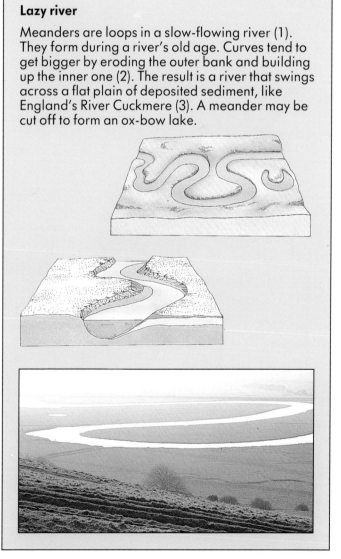

Weathering

Deltas

A river ends at its mouth. If there are few sea currents at that point, the debris builds up to form a delta. The delta can build out as banks along a series of channels, as in the Mississippi River (above and 1). Or it can spread out to form a semicircular area of triangular islands, as in the deltas of the River Nile (2) and the River Niger (3).

1 Mississippi

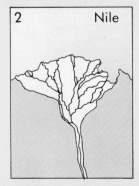

2 Nile

3 Niger

Much of the debris that falls into rivers and gets washed away has been broken from exposed rocks. This has been done by the erosive action of both wind and rain, water and ice.

Rainwater, even without the addition of pollution, is quite acidic. As moisture condenses from the clouds and falls in the form of drops, it dissolves carbon dioxide gas from the atmosphere. It then becomes weak carbonic acid. On soaking into the ground, the acid reacts with particular minerals in the rocks. This is very noticeable in granite areas, where the feldspars in the granite are attacked and turned into soft clay minerals. The other minerals in the granite – quartz and mica – then fall loose and are carried away as sand. This is why many granite areas have china clay pits and dazzling white quartz beaches.

The other type of rock mainly affected by the acidity of rainwater is limestone. Limestone is composed almost entirely of the mineral calcite, which dissolves in weak acids. The water flowing through the rock dissolves it to form galleries and caverns. Within them, the water redeposits the dissolved calcite as stalactites and stalagmites. The calcite is also redeposited in kettles and water pipes, giving the problem known as hard water.

The rocks of dry areas also have trouble with the weather. The wind in the desert can pick up sand particles and hurl them against exposed rocks and cliff-faces. This wears them away in a natural "sandblasting" effect, and produces yet more sand that can erode more rock. The infrequent rains that do fall in the desert soak into the surface layers of exposed rocks. This breaks down the minerals near the surface, and weakens the outermost skin. During the hot days and cold nights these surface layers expand and contract, and eventually split away to produce what is called onion-skin weathering. The rock comes apart layer by layer and leaves the core in the form of a rounded hill called an inselberg.

Features of a cave

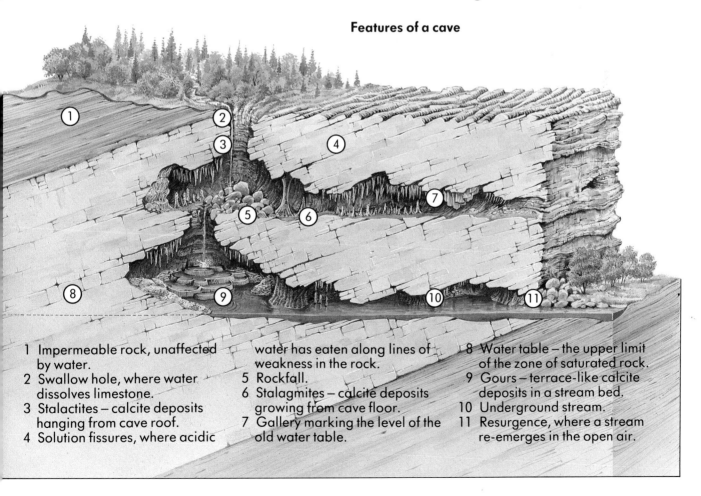

1 Impermeable rock, unaffected by water.
2 Swallow hole, where water dissolves limestone.
3 Stalactites – calcite deposits hanging from cave roof.
4 Solution fissures, where acidic water has eaten along lines of weakness in the rock.
5 Rockfall.
6 Stalagmites – calcite deposits growing from cave floor.
7 Gallery marking the level of the old water table.
8 Water table – the upper limit of the zone of saturated rock.
9 Gours – terrace-like calcite deposits in a stream bed.
10 Underground stream.
11 Resurgence, where a stream re-emerges in the open air.

43

Ice and frost

An ever-present problem in the northern winter is the possibility of water pipes freezing. When they do, they crack and split because of the expansion of the ice inside. Exactly the same thing happens in nature. On icy mountains water soaks into pores and cracks in exposed rocks. When it freezes, the water turns to ice. The ice expands with a pressure that widens the pores and cracks, splitting the rocks apart and breaking up entire mountainsides. Masses of broken blocks that slope downwards from craggy peaks are a result of this destructive action. They are called scree slopes.

A similar mechanism brings stones to the surface of a garden in winter time. Water beneath a buried stone freezes more readily than that in the soil around because the stone absorbs its heat more quickly. Ice forming beneath the stone expands and pushes the stone upwards. In permanently cold regions the whole soil surface is raised in a regular series of low humps. The stones brought to the surface collect in the troughs between the humps and produce a honeycomb-like pattern.

Ice can break and change rocks. It can also transport things. A glacier is a mass of ice that moves slowly under the influence of gravity. It is one of the strongest agents of transportation that there is. Much of the landscape of the Northern Hemisphere is formed from debris deposited by continent-wide glaciers during the ice ages of the last two million years.

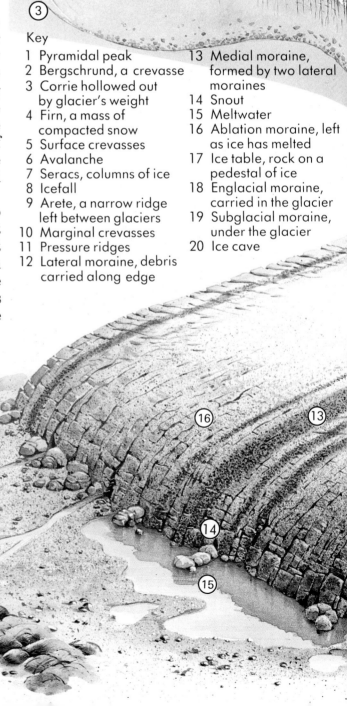

Key

1 Pyramidal peak
2 Bergschrund, a crevasse
3 Corrie hollowed out by glacier's weight
4 Firn, a mass of compacted snow
5 Surface crevasses
6 Avalanche
7 Seracs, columns of ice
8 Icefall
9 Arete, a narrow ridge left between glaciers
10 Marginal crevasses
11 Pressure ridges
12 Lateral moraine, debris carried along edge
13 Medial moraine, formed by two lateral moraines
14 Snout
15 Meltwater
16 Ablation moraine, left as ice has melted
17 Ice table, rock on a pedestal of ice
18 Englacial moraine, carried in the glacier
19 Subglacial moraine, under the glacier
20 Ice cave

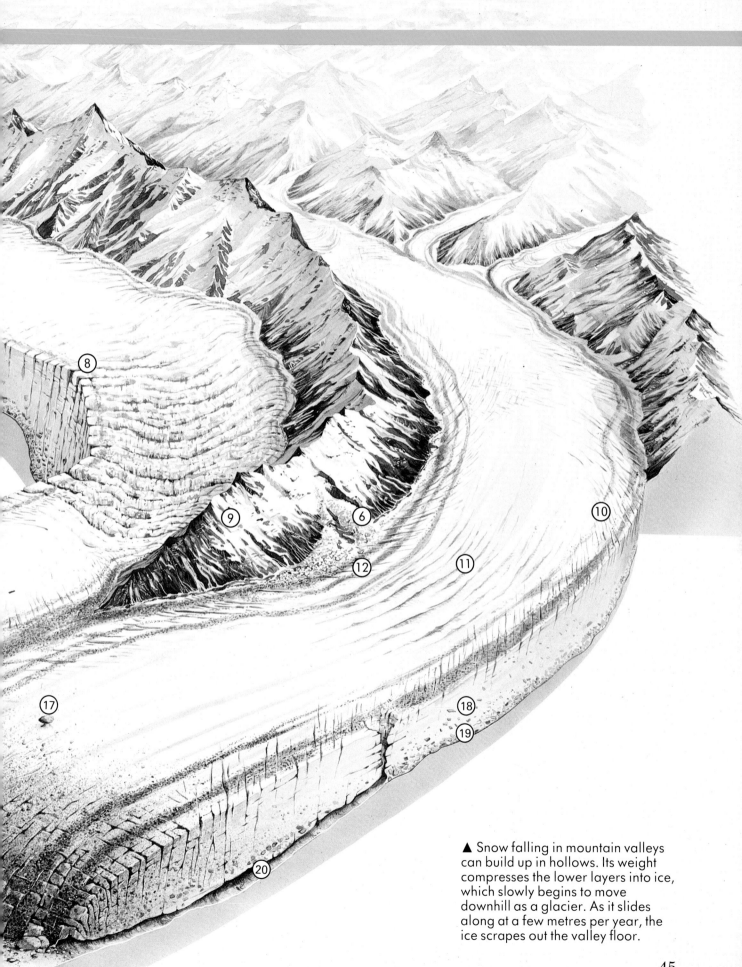

▲ Snow falling in mountain valleys can build up in hollows. Its weight compresses the lower layers into ice, which slowly begins to move downhill as a glacier. As it slides along at a few metres per year, the ice scrapes out the valley floor.

The changing scene

The landscape is never static. It changes from day to day, year to year, millenium to millenium. Most of these changes are natural. But with the coming of civilization, many have been caused by the activity of people.

Almost everything that civilization does has an impact on the surface of the Earth. About 5,000 years ago irrigation was first practised in the Middle East. Rivers were diverted to make dry desert areas fertile. The first cities were built, and these were often constructed on artificial hills, for defence. The people of low-lying lands by shallow seas often extend their farmland by walling off areas of sea and draining them. The Dutch have been doing this since the 7th century. Since the Industrial Revolution of the 18th century, vast areas of the landscape have been dug up for raw materials.

More spectacular are the unintentional changes in the landscape caused by human activities. Building breakwaters for harbours, or removing beach shingle for building, can alter sea currents. As a result, local seafronts can be washed away. Bad farming practices can alter the structure of the soil so that it falls apart and is washed away or blown away in the wind. In this way, fertile farmland can turn to desert very quickly.

With the population increasing every year, the impact that civilization has on the environment becomes greater and more significant all the time.

▼ The world's biggest hole in the ground is the Bingham Canyon Copper Mine in Utah. It has a depth of 774 m and covers an area of 7.21 sq km.

46

▲ In deserts, sand is blown along close to the ground at a height of about 1 m. It hits the base of rocks, eroding them into mushroom shapes.

► The sea is constantly eating away at the coastline. It undercuts cliffs and erodes headlands into isolated pillars of rock called stacks.

Igneous landforms

The shape of the landscape depends mainly on the type of rock it is made from. When an igneous rock fills a large crack, the structure is called a dyke. As the surrounding rocks are worn away, the dyke juts up like a wall across the scenery. An igneous rock forming a layer between beds of a sedimentary rock forms a structure called a sill. When this is eroded, it may look like a thick, hard sedimentary bed.

Sometimes the magma rising up a crack stops at a particular level and domes up the sedimentary rocks above it. This produces a laccolith, sometimes seen on the surface as a ring-like structure in the uplifted sedimentary rocks. When an ancient volcano is worn away, the solid igneous material in the vent may remain standing as a pinnacle (below), showing where the volcano once stood.

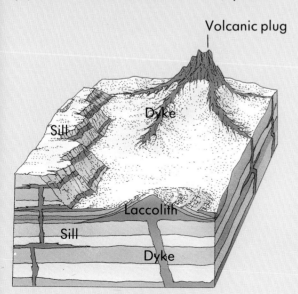

Volcanic plug

Dyke

Sill

Laccolith

Sill

Dyke

The oceans

Spot facts

• The oceans have a total area of 361,300,000 square kilometres. This represents 71 per cent of the total surface area of the Earth.

• The total volume of the oceans is 1,349,929,000 cubic kilometres.

• The deepest point is in the Marianas Trench, 11,033 m deep.

• Traces of all the chemical elements are found in the oceans, including 60,000 tonnes of gold.

• If removed from the sea, the dissolved salt would cover the land to a depth of more than 150 m.

We call our planet the Earth. It would be more appropriate if we called it the Water, because more than two-thirds of its area is covered by the seas and oceans. This is well brought out in satellite photographs and the views that astronauts have from space. The overall colour is blue, caused both by the effect of the atmosphere and by the vast areas of water beneath. It is the presence of all this water that has made life possible. All life processes involve water, and life evolved in the sea about 3,500 million years ago, and only about 500 million years ago did life move on to land.

► The constant movement of the ocean is seen in the pounding surf of the shoreline. The wind moves the waves. On a larger scale the winds and the Sun create the great ocean currents. The gravity of the Moon and Sun gives rise to the tides.

The waters

The water in the sea tastes very salty. Drinking it is likely to make you sick. Seawater is not pure, but contains a large quantity of salts that have been dissolved from the rocks of the Earth's crust. The actual salinity, or saltiness, varies from place to place. It is saltiest in warm enclosed seas, where water is constantly being evaporated. It is least salty in the cold northern and southern oceans, where it is diluted by rain and melting ice. It is also less salty where great rivers like the Amazon or Niger flow into the ocean. However, the proportions of the different salts present are constant throughout the world.

The temperature of seawater varies a great deal over the surface of the ocean, but the temperature beneath is fairly constant. The slightest variation in temperature can trigger ocean currents.

Combinations of salinity, temperature, currents and other factors determine the amount of life present in any ocean region. Vigorous life is confined to the surface and a few hundred metres below it. There the Sun shines into the water and plant life, some of it microscopic, can grow. Small invertebrates feed on the plants, and fish feed on these. A whole ecosystem is supported. In the dark depths the Sun has never shone. The only life consists of creatures that feed on dead organic debris, which rains down from the more fertile layers above, or that prey on one another.

▼ The distribution of land and sea on our globe is not even. Most of the ocean area lies as the Pacific on one side. This is an accident of plate tectonics. 200 million years ago the inequality was even greater, with all the continents fused together and the rest of the world covered with a single ocean called Panthalassa.

Seawater composition

Much of the gas belched out of volcanoes is water vapour. It is actually recycled seawater, which has been carried down into the depths of the crust by plate tectonic movement. The salinity of ocean water varies usually between 33 and 38 parts per thousand. The dissolved salts contain nearly all the chemical elements. The commonest are sodium and chlorine, which together make up common salt.

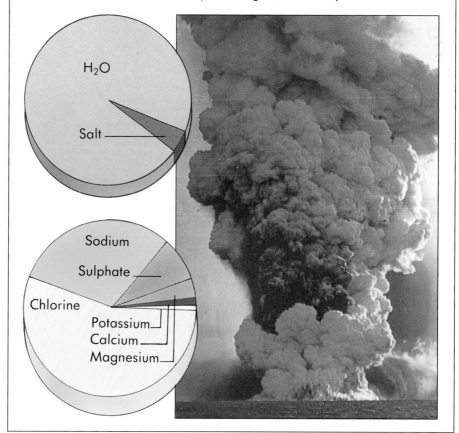

49

The ocean surface

The continents are large islands, mostly quite separate from each other. As a result, the oceans form a single continuous body of water. In some places, such as the Bering Strait, the gaps between the continents are narrow; in others, such as the East Indies, they are choked with islands. But nowhere is there an area of ocean that is completely isolated from any other. Political problems aside, we can sail from a port on one continent to any other seaport in the world. For convenience, however, we often talk about the seven seas. These are the North Atlantic, South Atlantic, Indian, North Pacific, South Pacific, Arctic and Antarctic Oceans.

The surface of this vast waterway is constantly in motion, notably through the movements of ocean currents. Vertical currents are set up by convection. This happens because a cold mass of fluid is denser than a warmer mass of the same fluid and sinks through it. Cold water from melted ice at the North Pole sinks through the surrounding warmer water. It then travels for thousands of kilometres as an undersea current along the bed of the Atlantic. On the surface, however, it tends to be the prevailing winds that power the ocean currents. The Trade Winds are those that blow towards the Equator from the north-east and south-east. The surface waters blown by them produce an overall westward flow of equatorial water as the equatorial currents.

When this water reaches a continent, it sweeps north and south, producing warm currents along the east coast of that continent. The Gulf Stream in the North Atlantic, the Kuro Shio in the North Pacific and the Australian Current in the South Pacific are examples. The movement is completed when the water sweeps towards the Equator along the eastern edge of the ocean and joins up with the beginnings of the equatorial currents. These include the Californian Current and the Peru Current in the eastern Pacific Ocean.

Hence, the world pattern of ocean currents is based on a vast system of circular movements, or gyres, each occupying half an ocean. The movement of warm currents along cold continental edges, and cold currents along warm continental edges, helps to modify the climate in these coastal areas.

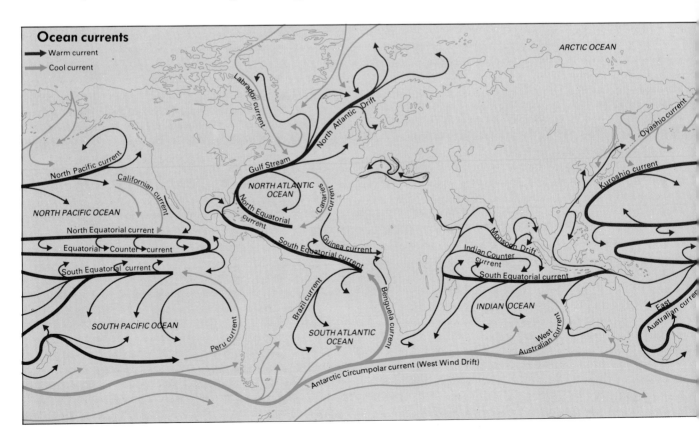

Ocean currents
Warm current
Cool current

ARCTIC OCEAN
Labrador current
North Atlantic Drift
Oyashio current
North Pacific current
Californian current
Gulf Stream
NORTH ATLANTIC OCEAN
Canaries current
Kuroshio current
NORTH PACIFIC OCEAN
North Equatorial current
North Equatorial current
Equatorial Counter current
Guinea current
Monsoon Drift
Indian Counter current
South Equatorial current
South Equatorial current
South Equatorial current
Brazil current
Benguela current
INDIAN OCEAN
SOUTH PACIFIC OCEAN
Peru current
SOUTH ATLANTIC OCEAN
West Australian current
East Australian current
Antarctic Circumpolar current (West Wind Drift)

◄ Penguins and flat-topped icebergs typify the cold Antarctic Ocean. The icebergs are made from glaciers that have flowed seawards off the edge of Antarctica. The ice at the North Pole, on the other hand, is formed on the sea's surface.

◄ Prevailing winds drive the ocean currents. The Westerlies blowing in the far south produce the cold circumpolar current called the West Wind Drift. This separates the Antarctic from warmer waters farther north and keeps Antarctica frozen. Elsewhere, the ocean gyres bring warm or cool currents to the edges of the continents.

► During the ice ages of the last 2 million years the sea level changed. So much water was locked up in the expanded ice cap at the North Pole that the volume of liquid water left was less than it is now and the sea level was lower. Land appeared where there is now shallow sea. At the same time in the far north, the great weight of the ice on the land pushed the land downwards, and sea levels were higher than they are today. The ice came and went several times, and we can often see the different sea levels as "raised beaches". These are banks a few metres above sea level that mark the ancient shorelines.

Today

An ice age

The ocean floor

Under the sea is a varied landscape that is rarely seen by human beings. It is a landscape of mountains, volcanoes and broad, flat plains. Only since the 1960s have we really begun to understand it.

There are a number of different zones of the ocean floor. First there is the continental shelf. This is merely the edge of the continent that is awash. If we drill through the sediment on the continental shelf, we find continental crust beneath it, not oceanic crust. The continental shelf can be hundreds of kilometres broad where it is at the edge of an old continent, as for example in the North Sea and Hudson Bay. But the shelf is very narrow or non-existent where new mountain ranges are being pushed up along a coastline, as along the western coast of South America.

At the edge of the continental shelf the seabed slopes downwards into the depths. This feature is known as the continental slope, an it marks the edge of the continent itself. In some places – usually offshore from the mouths o large rivers – the continental slope is cut by vas canyons, wider and deeper than any canyon or land. These have been eroded out by debris swept down from the rivers.

At the foot of the continental slope lies the continental rise. This is not as steep as the

▼ Coral reefs are found only in shallow tropical waters. Coral builds out from an island in a shelf called a fringing reef. Over thousands of years the island may sink as the coral builds up. A lagoon then separates the reef, now a barrier reef, from the dwindling island. When the island sinks completely, it forms an atoll, which is a ring of coral surrounding a lagoon.

continental slope itself and is built up from debris that has been brought down from the continental shelf above. Perhaps the largest are the fans of material that spread out through the northern Indian Ocean from the outflowings of the rivers Ganges and Indus.

The abyssal plain is the floor of the ocean itself. With true oceanic crust as its foundation, it extends between the continents. It is covered with a fine muddy deposit called ooze. This is made up of the remains of millions of sea creatures that have settled on the dark bottom over many millions of years. There is no debris from the land here. Across the abyssal plain rise the volcanic ocean ridges. Occasionally the plain drops away into the ocean trenches, the deepest points of the ocean, where the old crust is continually being destroyed.

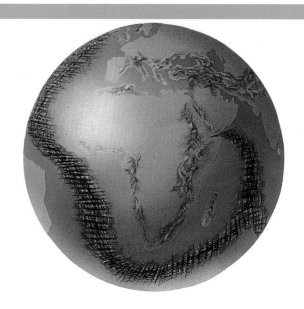

▲ The ocean ridges are the most extensive geographical feature on the Earth. They run through the beds of all the oceans without a break.

Geography of the ocean floor

The continental shelf consists of continental crust and reaches out from the shore to a depth of about 150 m. At its outer edge it drops away as the continental slope, with a gradient of between 3 and 20 degrees. The slope, and the continental rise at the bottom, gives out on to the abyssal plain at a depth of about 4,000 m. Ocean ridges rise from the abyssal plain to heights of between 500 and 1,000 m. Volcanoes are constantly erupting in the rift valley that exists along its crest. The ridge volcanoes may occasionally reach the ocean surface, but more often they are completely submerged. Seamounts on the abyssal plain, some with flat eroded tops, are the remains of these volcanoes. The ocean trenches, in which one part of the Earth's crust is being swallowed up beneath another, are the deepest points on Earth.

▼ The volcanic activity in an ocean ridge is dramatically shown by the presence of "smokers". These are jets of hot water that burst from the Earth's crust at great depths. The "smoke" is caused by fine particles of minerals in the water.

The ocean floor

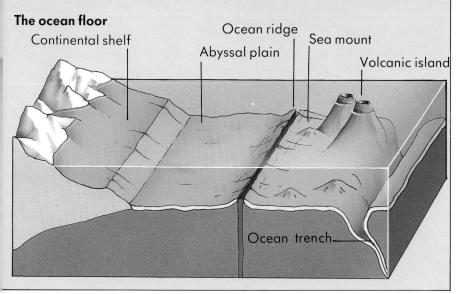

Continental shelf
Ocean ridge
Sea mount
Abyssal plain
Volcanic island
Ocean trench

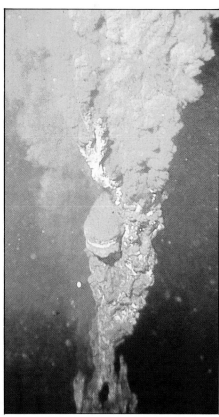

Cycling the water

More than two-thirds of the Earth's surface is covered with water. By far the greatest volume of this – 97.2 per cent – is contained in the oceans. The remainder is found as water vapour, fresh water and ice. Water occurs as vapour in the atmosphere, as ice in the glaciers and ice caps, and as liquid water flowing in rivers, standing in lakes and swamps, and absorbed into rocks and soil as groundwater. Water vapour in the atmosphere condenses to form raindrops, which fall to the ground. Rainwater collects to form streams and rivers, which flow to the sea. If it were not for water, which is available in most places on the Earth, life would be impossible on our planet.

Moving water

Physical conditions on the Earth's surface ensure that water can exist in all three of its possible states. It can exist as a gas – as water vapour; it can exist as a liquid – in the form that we most often see it; or it can exist as a solid – in the form of ice. It takes a relatively small shift of conditions such as temperature or pressure to change water from one state to the next. One cold night can turn a liquid pond into solid ice, or one hot day can evaporate a large puddle to vapour, leaving the surface dry.

Water is continually evaporating into the atmosphere from open bodies of water, such as seas and lakes. Then it condenses and falls as rain. The rain that falls on the land runs off the surface or sinks into the rocks and soil to form the groundwater. It re-emerges at springs and forms the beginnings of streams and rivers, and flows back to the lakes and oceans. This whole process is known as the water cycle.

The cycle has many side branches as well. Water that falls on the ground may evaporate straight away. Groundwater is drawn up through plants and evaporated from leaves.

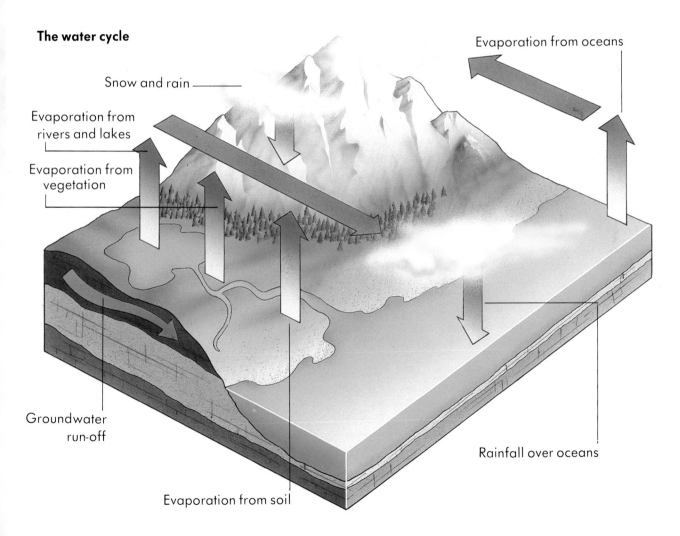

The water cycle

Evaporation from oceans

Snow and rain

Evaporation from rivers and lakes

Evaporation from vegetation

Groundwater run-off

Rainfall over oceans

Evaporation from soil

▲ Water goes round and round in the water cycle. The main upward movement is by evaporation. Water evaporated as vapour from oceans, rivers and lakes passes into the atmosphere. Some water vapour is also "breathed out" by plants. High in the air, water vapour condenses to form clouds. Sideways movement occurs when clouds are blown along by the wind. Finally, downward movement, to complete the cycle, takes place when it rains. When it is cold, rain falls as snowflakes or hailstones, and lies as snow and ice. Sometimes snow and ice can even change directly into vapour without passing through a liquid phase.

Water resources

We all need water to keep us alive. Each of us needs at least 2 litres every day just to keep the body working. In Westernized societies the daily consumption increases because of the water used for washing and the demands of industry and agriculture. On average it takes about 2,000 litres of water a day to support one person in an industrial nation. It is no wonder that water is regarded as one of the most valuable of the Earth's resources.

The Earth's surface is largely covered by water, but unfortunately most of this is unusable or in the wrong place. Many rapidly expanding centres of population lie in areas where there is very little water because the climate is too dry and the rainfall irregular. Other expanding centres of population lie where the presence of water is an embarrassment, such as by tropical rivers that are likely to flood and promote the spread of disease.

Much of the usable water is obtained from rivers. The irregular or seasonal flow of a river can be regulated and controlled by using dams.

Built across the flow of a river, a dam traps the water and fills the valley behind it to produce a reservoir. The larger a dam becomes, however, the more problems are associated with it. A heavy structure on unstable ground could collapse, and with several million tonnes of water behind it the result would be disastrous. Reservoirs may also silt up, because the original flow of the river is disrupted. The floor of the reservoir builds up so that it becomes shallow and can hold only a small fraction of the original water. The stagnant surface of the water may also promote the growth of water weed that both chokes waterways and makes the water unmanageable.

The other great water source is groundwater,

▶ The Kariba Dam, at the border between Zambia and Zimbabwe, in southern Africa. The dam creates the large reservoir of water called Lake Kariba.

▼ The plain of the River Indus in India is irrigated by a complex system of channels that bring the spring meltwater from Himalayan glaciers to the fields.

the water that has soaked into the rocks and soils of an area. This can be extracted by digging wells or boreholes and installing pumps. In the driest areas, however, this does not represent an inexhaustible supply. Much of the water that is being pumped up in northern Africa is actually rain that fell in the Ice Age. Once that has gone, there will be no more.

As the world population grows there will be bigger and more elaborate schemes for distributing water to where it is needed. Each will bring its own problems.

The constant cycling of water is well seen in the daily downpours that occur in the areas of tropical forest that straddle the Equator.

Groundwater

Springs, wells and oases

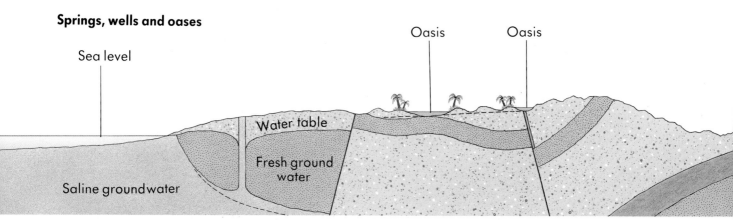

Sea level

Oasis

Oasis

Water table

Fresh ground water

Saline groundwater

When rain falls on the soil, some of it washes off as surface flow. The remainder sinks into the soil and becomes groundwater.

Soils tend to be quite loose and have large air spaces and pockets between the particles. These are easily filled up with water, in a similar way to the holes in a sponge. The rocks beneath are more compact, but usually contain air pores as well. If the pores are connected together, water can soak through. We say that the rock is permeable. At some distance below the surface the rocks are so compact that they have no spaces in them and water cannot penetrate. The rock is impermeable. When the rock above this level is full of water, it is said to be saturated. Above the saturated zone is a region in which the water is seeping downwards. This is the zone of intermittent saturation.

The upper boundary of the saturated zone is called the water table. Its level varies from time to time, being higher in wet weather and very low in times of drought. When wells are drilled, they are driven down to below the water table. Water from the surrounding rocks gathers in the bottom of the hole. Where the water table reaches the surface, the water leaks out and becomes a spring. Other springs form when the water from the saturated zone seeps upwards through cracks or faults.

The movement of water through underground rocks is slow, usually taking years. Once underground, the water is protected from evaporation, and its passage through the pores of the rocks filters it so that it is usually quite clean. Accordingly, underground water is a valuable resource, although there may be some salts and minerals dissolved in it.

It is estimated that about 62.5 per cent of the world's fresh water is present in the form of groundwater in the rocks.

◀ Groundwater is widely exploited in areas where there is little rainfall and few rivers. In desert oases, such as this one at Taghit in Algeria, the water table reaches the surface and a freshwater lake forms. The water is used for drinking and irrigation.

▶ In most instances the water table lies well below the surface, and the water has to be brought up by artificial means. Here oxen are used to turn a wooden gear that brings water to the surface by a bucket wheel for distribution to the groves of date palms. Desert peoples have become skilful at drawing water and transporting it to the fields by irrigation systems.

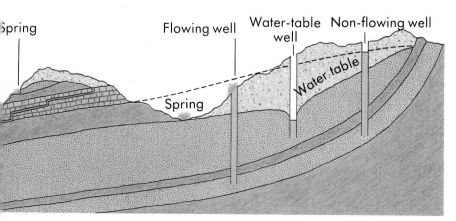

Spring   Flowing well   Water-table well   Non-flowing well

Spring

Water table

◀ A rock that contains groundwater is called an aquifer. Different kinds of wells can be drilled to reach water in an aquifer. A particular type is the artesian. Exposed rocks absorb water in a hilly area. In the nearby lowlands a well drilled through the overlying impermeable bed into the aquifer reaches the water. The pressure of the water coming down the slope of the aquifer pushes the water up the borehole until it gushes out at the surface.

Clouds and rain

The water vapour in the atmosphere makes up only about 0.001 per cent of the total water supply of the world. But without it all life would be confined to the sea. The proportions of most of the gases in the atmosphere are constant wherever we go, but the proportion of water vapour varies.

The concentration of vapour in the atmosphere depends largely on the temperature and pressure. When the atmosphere is holding as much vapour as it can at a particular temperature and pressure, it is said to be saturated. Any change to these conditions will turn some of the vapour back to liquid. Because the atmosphere is always in motion, the physical conditions are changing all the time. So is the amount of water in the air.

Such changes can be caused by the wind blowing from a warm region to a cold region. Convection currents may lift warm air up to heights where the pressure is less. Wind blowing off the sea can carry moisture up a hillside, where the air becomes cooler as it rises. When these things happen, the vapour in the air turns to water in the form of tiny droplets. These are too light to fall to the ground and remain suspended as clouds.

Different types of clouds form under different conditions. At very great heights, above about 7,000 m, the conditions produce clouds of ice crystals rather than droplets. At other heights the water droplets form layer clouds, called stratus; or heaped clouds, called cumulus; or a combination of the two. Thunderclouds have such strong convection currents that the main cloud mass is made of water droplets, while its crown is of ice.

Cooling and decreasing the pressure still further causes the droplets to mass together into larger drops, and these fall as rain.

▼ A satellite view of a hurricane. The rapid heating of surface ocean waters in tropical regions causes hurricanes to form. Turbulent and constantly changing conditions are found in the hurricane. Winds spiral in towards a low-pressure area and thick banks of clouds form as the pressure becomes less and less. Torrential rains fall as the pressure reduces still further.

▶ Thunderstorms develop where a patch of air heats up quickly and rises. Raindrops form under the reduced pressure at height, and are carried up and down by the strong currents. The growing drops become too big to be stable and split into smaller drops, producing an electrical discharge. A strong charge builds up in the cloud and flashes to earth as lightning.

▲ Heavy seasonal rains called monsoons develop around the north of the Indian Ocean. In the winter the Asian landmass cools and dry air flows seawards. In the summer the continent heats up and draws in wet air from the ocean. The moisture falls as rain over the land.

▼ Snow is frozen water. At certain temperatures the vapour in the atmosphere does not form water droplets, but forms ice crystals instead. When the crystals clump together and fall to the ground, the result is snow. When raindrops freeze, they fall as hail.

Life-giving atmosphere

Spot facts

● The atmosphere weighs about 5,000 million million tonnes.

● At sea level, the weight of the atmosphere exerts a pressure of 1.05 kg on every square centimetre.

● We can study the change in climate over the last half million years by looking at the composition of the ancient ice in Antarctica and Greenland.

● The greenhouse effect may help to avert another ice age.

The Earth's atmosphere is an ocean of air that surrounds the planet. Air is a mixture of gases, mainly nitrogen and oxygen. We cannot see it, or taste it or smell it, yet air is vital for life. Layers of gases high in the atmosphere shield us from harmful radiation from the Sun. But the increasing destruction of these protective layers caused by pollution is altering the Earth's climate. We take oxygen from air into our bodies with every breath, and without it we suffocate. Oxygen in the air is also needed for fuels to burn. And air does have substance. Without it, birds and aircraft could not fly.

▶ This beautiful sunset was caused by dust in the lower atmosphere. Dust scatters blue light, but lets red light pass through.

The air we breathe

The atmosphere is a thin layer of air that surrounds our planet. On a small model of the Earth its thickness would be hardly noticeable – no thicker in than the skin of an apple. But its weight at ground level gives us the air pressure under which we and all other land-living creatures evolved. The higher up we go, the thinner the atmosphere becomes. At a height of several hundred kilometres it fades away into the vacuum of space.

The atmosphere is a mixture of gases. The main gases are nitrogen and oxygen, with about four parts of nitrogen to every one of oxygen. The remainder – just over one per cent of the whole – consists of carbon dioxide and rare gases like argon, helium and neon. These rare gases are called noble or inert gases because they do not take part in chemical reactions. The proportions of all these gases tend to remain the same all the time.

▶ Most of the atmosphere consists of nitrogen, but about 21 per cent of it is oxygen generated by the action of plant life. The oxygen makes Earth's atmosphere different from those of other planets.

▼ Mountaineers who climb to high altitudes have to carry a supply of oxygen to breathe. At high altitudes the air pressure is less, and so there is less oxygen available n the air.

There are, however, a few variable components in the atmosphere. The most important variable is water vapour. It can be almost absent in desert areas but can reach a concentration of about three per cent in very humid regions. The presence of water vapour is essential to life. Sulphur dioxide is another variable. This is not essential to life, and can in fact be quite harmful. It is produced in large amounts by volcanoes and by burning fuels such as coal and oil.

At a height of between 15 and 50 km there is the so-called ozone layer. Ozone is a type of oxygen. High-energy ultraviolet radiation from the Sun affects oxygen in the layer and turns it into ozone. This reaction absorbs ultraviolet radiation and prevents most of it from reaching the Earth's surface. The layer is important because too much ultraviolet radiation would be harmful to living things.

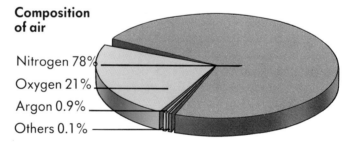

Composition of air

Nitrogen 78%
Oxygen 21%
Argon 0.9%
Others 0.1%

The atmosphere's structure

The layers of air that form the atmosphere stretch upwards above our heads for about 700 km. At that height the air is extremely thin. At even higher altitudes the atmosphere fades away into the airless vacuum of space.

The "thickness" of the atmosphere is called its density. Air is densest near the surface of the Earth. It is less dense at the tops of tall mountains. Atmospheric pressure is also greatest near the ground. The pressure is caused by the weight of layers of air pressing down from above.

The atmosphere can be divided into a number of layers, each with its own properties. The lowermost 11 km or so is called the troposphere. Although it is quite a thin layer, the troposphere contains, under pressure, the greatest proportion of air by mass. All the physical activities that affect the weather take place in this region.

The top of the troposphere is a theoretical boundary called the tropopause. Above this lies the stratosphere, extending up to about 50 km. Most military and long-distance aircraft operate in this region. The ozone layer, within this region, absorbs much of the Sun's energy. As a result, the temperature is quite high in the stratosphere.

Above the stratopause – the upper limit of the stratosphere – stretches the mesosphere, up to about 80 km. The temperature there is low and the air is thin. But it is still thick enough for meteorites to burn up as they pass through it.

Beyond its upper boundary, the mesopause, the mesosphere gives way to the thermosphere. This is a another region of high temperatures caused by absorption of solar radiation. Then comes the exosphere, which eventually fades away to nothing at about 700 km above the surface of the Earth.

▶ Layers of the atmosphere. The weight of the air pressing down on itself compresses the lower layers. As a result the lowermost layer, the troposphere, contains 80 per cent of the atmosphere by mass. But it occupies a volume of only 1.5 per cent. Above the stratosphere there is only one per cent of the mass of air, but this is spread through 93 per cent of the volume. The two diagrams (right) compare the composition of the atmosphere in terms of mass and in terms of volume.

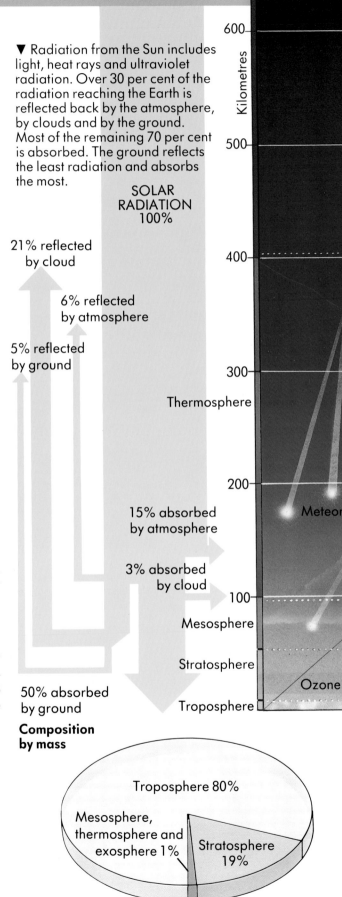

▼ Radiation from the Sun includes light, heat rays and ultraviolet radiation. Over 30 per cent of the radiation reaching the Earth is reflected back by the atmosphere, by clouds and by the ground. Most of the remaining 70 per cent is absorbed. The ground reflects the least radiation and absorbs the most.

SOLAR RADIATION 100%

21% reflected by cloud

6% reflected by atmosphere

5% reflected by ground

15% absorbed by atmosphere

3% absorbed by cloud

50% absorbed by ground

Composition by mass

Troposphere 80%

Mesosphere, thermosphere and exosphere 1%

Stratosphere 19%

700
Exosphere

600

Kilometres

500

400

Thermosphere

300

200

Meteori

100

Mesosphere

Stratosphere

Ozone l

Troposphere

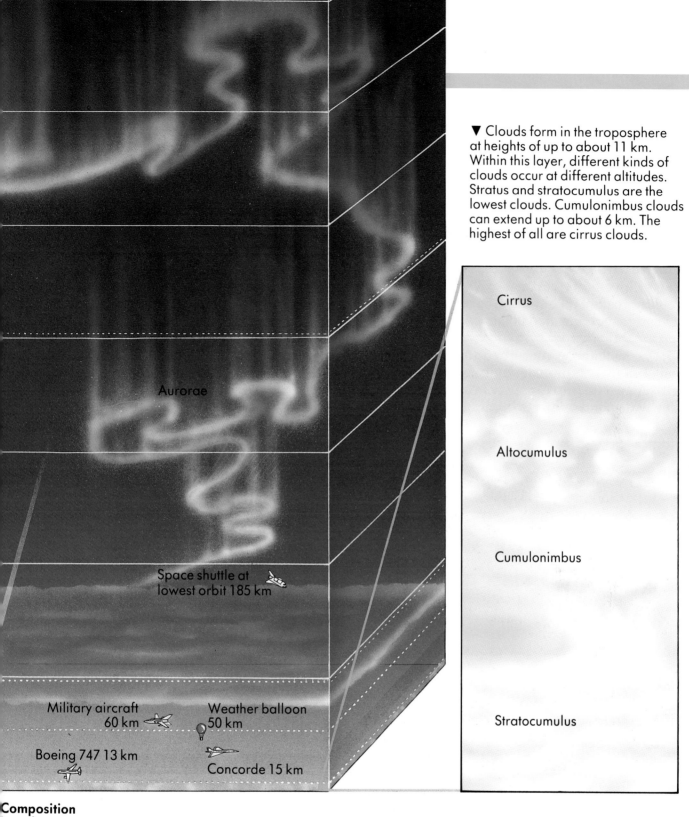

▼ Clouds form in the troposphere at heights of up to about 11 km. Within this layer, different kinds of clouds occur at different altitudes. Stratus and stratocumulus are the lowest clouds. Cumulonimbus clouds can extend up to about 6 km. The highest of all are cirrus clouds.

Aurorae

Space shuttle at lowest orbit 185 km

Military aircraft 60 km

Weather balloon 50 km

Boeing 747 13 km

Concorde 15 km

Cirrus

Altocumulus

Cumulonimbus

Stratocumulus

Composition by volume

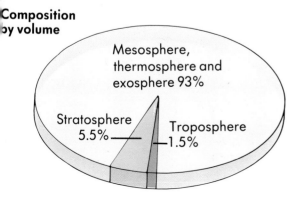

Mesosphere, thermosphere and exosphere 93%

Stratosphere 5.5%

Troposphere 1.5%

▲ The different layers of the atmosphere have different properties. The interaction between the solar wind and the Earth's magnetic field produces the light effects of the aurora in the exosphere and the thermosphere. About half of the solar radiation is absorbed or reflected by different layers before it reaches the ground. This absorption produces different temperature conditions in different layers. Meteorites burn up in the upper layers. Only the troposphere has enough oxygen to support life.

The changing atmosphere

The atmosphere that formed as the Earth first solidified was very different from the atmosphere of today. Even now it gradually continues to change.

The first atmosphere probably consisted mostly of carbon dioxide, nitrogen, hydrogen, carbon monoxide and inert gases. The strong solar wind would immediately have blasted much of this away into space. Then as the Earth began to solidify, gases were emitted from the cooling rocks and built up the next atmosphere. These gases consisted largely of carbon dioxide, with some nitrogen, hydrogen and traces of argon. Volcanoes continued to bring up water vapour, carbon dioxide, hydrogen sulphide and nitrogen. The Sun's energy broke down some of the water vapour into hydrogen and oxygen.

It also converted some of the oxygen into ozone to produce an ozone layer early in Earth's history. The first oceans formed when so much water vapour was released that the atmosphere could not hold it all. The vapour condensed into clouds, and it began to rain.

The next major change in the atmosphere took place when carbon dioxide dissolved in the early oceans. We can tell that this was happening because we can find early rocks that contain calcite. This mineral was formed from carbon dioxide dissolved in sea water. The level of carbon dioxide in the atmosphere fell from about 80 per cent to its present level of about 0.05 per cent by about 1,000 million years ago. Meanwhile, the hydrogen in the atmosphere was leaking off into space. It was too light to be held firmly by the Earth's gravity. As a result of the loss of these gases, the proportion of nitrogen gradually grew until it reached its present proportion of about 80 per cent.

Enter oxygen

The most important change to the atmosphere began about 2,500 million years ago. Before this time there was very little oxygen in the atmosphere. We can tell this because of the rocks that formed at the time. The iron in them formed minerals that were poor in oxygen. But if there had been free oxygen in the atmosphere, the iron would have formed rust-red

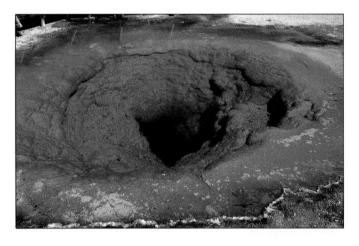

▲ This colony of green algae is living on chemicals in the hot, oxygen-free water of a hot spring. The first living things must also have lived in an oxygen-free environment. Their biochemistry was what scientists call "anaerobic".

► Colonies of aerobic bacteria. Many types of these live in stagnant lakes and estuaries. The colonies shown here are living in black mud that contain no oxygen.

▲ The burning of coal, oil and wood has an effect on the composition of the atmosphere. Burning uses up oxygen from the air and releases carbon dioxide. Water vapour is also produced by burning these fuels. Carbon dioxide and water vapour together give rise to what is known as the greenhouse effect. The Sun's rays pass through the atmosphere and warm the Earth in the normal way, but the excess heat cannot escape back out again. Sulphur dioxide is also produced by burning coal, and this reacts with the moisture in the air to produce sulphuric acid. The result is acid rain, which damages plants and poisons lakes wherever it falls.

minerals. The oldest red beds, with iron oxides, date from about 2,500 million years ago. At about this time primitive living things in the sea were beginning to use the energy of the Sun to make their food. They were the first plants. A by-product of this activity was the generation of oxygen. The oxygen gradually built up until reached its present level about 500 million years ago.

Now the atmosphere is changing again. Large-scale forest clearance cuts down the amount of oxygen produced, industry adds carbon dioxide to the atmosphere, and many processes disrupt the ozone layer. The long-term effects of these changes on climate and on living things have yet to be seen.

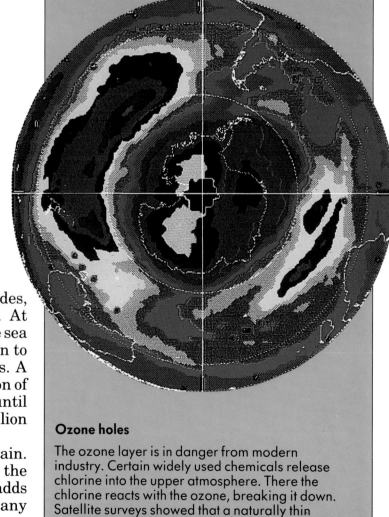

Ozone holes

The ozone layer is in danger from modern industry. Certain widely used chemicals release chlorine into the upper atmosphere. There the chlorine reacts with the ozone, breaking it down. Satellite surveys showed that a naturally thin portion of the ozone layer over the South Pole (shown in pink) was becoming bigger in the 1980s.

67

Weather and climate

Spot facts

● *Frozen rain can be carried up and down in a thundercloud until it forms large hailstones as big as tennis balls.*

● *Jet streams – narrow belts of high winds that circle the globe at great heights – can reach speeds of 200 km/h.*

● *Greenland acquired its name because of the plant growth produced by the warmer weather there in Viking times.*

● *Nuclear war could lead to a "nuclear winter", during which sunlight would not be able to penetrate the clouds of smoke and dust produced.*

▶ A hurricane is a violent storm of circling winds. A region of very low pressure develops over the ocean, and the air from nearby high pressure regions spirals in to balance it. This generates winds of up to 300 km/h. The centre of the hurricane moves and causes great damage wherever it passes.

There is a difference between weather and climate. In a particular geographical region, climate is the overall result of the atmospheric conditions, averaged over a long period of time. The conditions include temperature, atmospheric pressure and wind patterns. Weather, on the other hand, depends on the day-to-day variations of these conditions. Weather is important to agriculture, fishing, transport and many other human other activities. This is why meteorology – the scientific study of weather – is a very important science. Because so many people depend on them, weather reports have to be accurate.

Weather systems

The pattern of weather depends on the distribution of regions of warm air and cold air. This, in turn, depends on the distribution of low pressure areas and high pressure areas over the globe. When the Sun shines on the Earth's surface, the ground warms up. This warms the air above it. Warm air is lighter than cool air and so it rises, producing an area of low atmospheric pressure. In the cooler areas round about, the air is denser and at a higher pressure. It begins to move in towards the low pressure area. This movement produces the winds. The global wind pattern can be pictured as a reflection of the areas of high pressure and low pressure across the world.

▼ A front is the boundary between two air masses. The boundary is not stationary, but moves across the landscape. If a warm air mass is replaced by a cold one, then it is a cold front marked on a weather map by a line with teeth. A cold air mass is replaced by a warm one at a warm front represented by a line with semicircles. At a cold front the cold air wedges its way

When air rises it cools. But cool air can hold less moisture than warm air. As a result, moisture forms water droplets in a cooling air mass. These droplets form clouds and, ultimately, it rains.

As a general rule, wherever there are rising air masses there is rain. The air may be made to rise by convection currents, by winds blowing up a mountain range, or by a cold air mass pushing its way below a warm air mass where the two meet along a "front". Such fronts are found between the cold air over the North Pole and the warmer air over the tropics. They account for the unstable weather patterns in temperate Europe and North America.

beneath the warm, forcing the warm air upwards. Thunderclouds often form here as the warmer air is swept up quickly. At a warm front the warm air rises gently above the cold, producing a series of clouds at different heights. When two cold air masses catch up with each other and lift the intervening warm air, they form an occluded front .

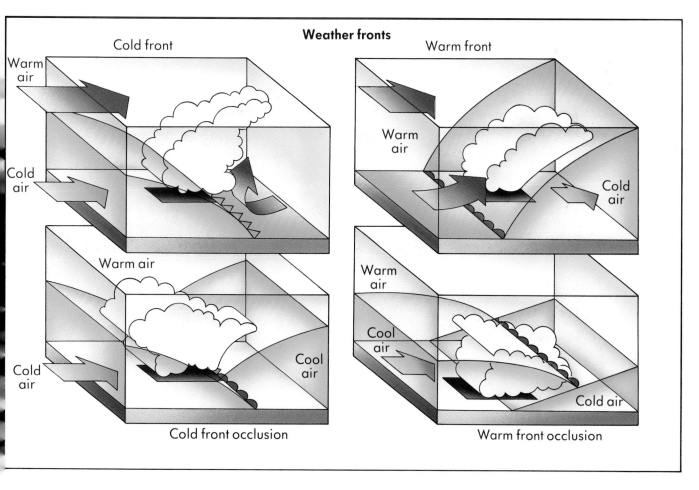

Weather fronts

Cold front

Warm front

Warm air

Cold air

Warm air

Cold air

Warm air

Cold air

Cold front occlusion

Warm air

Cool air

Cool air

Cold air

Warm front occlusion

Wind

Winds are produced when cool air moves in to replace the warm air rising in a low-pressure area. The world distribution of high and low pressures determines the pattern of the usual, or prevailing, winds.

The Sun is always directly overhead somewhere between the tropics. As a result, the land areas along the Equator are among the hottest on Earth. The hot air rises there, resulting in an equatorial low-pressure belt. Air from the north and south sweeps in to equalize the pressure. This gives rise to the North-easterly Trade Winds and the South-easterly Trade Winds. (Winds are always named after the direction from which they blow.) The north-south movement of air is deflected towards the west as a result of the turning of the Earth.

The warm air that rises at the Equator spreads northwards and southwards at the top of the troposphere. There it cools before descending again in the regions of the Tropics of Cancer in the north and Capricorn in the south. Tropical high-pressure belts form there. The world's greatest deserts are found along these belts, because the descending air is dry. The air that descends may then return towards the Equator as the Trade Winds. Or it may spread towards the more temperate regions as the warm South-westerlies in the Northern Hemisphere or the North-westerlies in the Southern Hemisphere. These winds are usually referred to simply as the Westerlies.

Over the North and South Poles the cold temperatures give rise to high-pressure regions of cold air. Cold winds spread outwards from these regions and meet the Westerlies along frontal systems in the temperate regions. Where they meet, the weather patterns are unstable. These major air movements produce the basic circulation pattern of the world's winds. It is disrupted and altered by the distribution of land and sea, and by the presence of mountain ranges. All of these factors decide the various climates of the world.

▼ Damage caused by a tornado in Pennsylvania in 1965. The tremendous power of the winds is spectacularly illustrated by the damage done by storms.

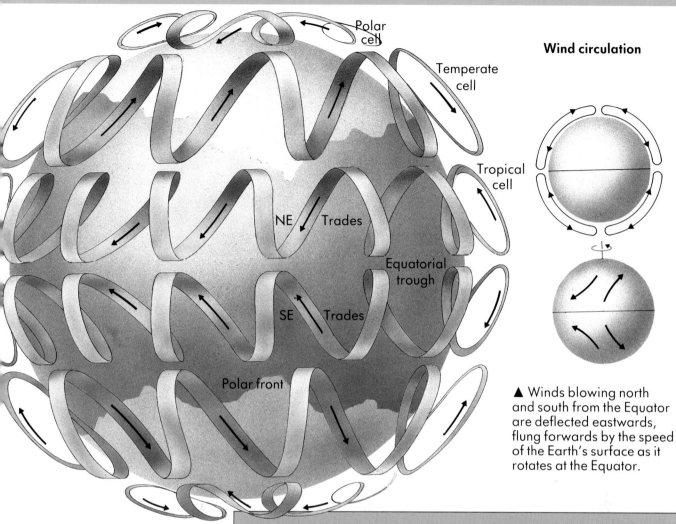

Polar cell

Temperate cell

Tropical cell

NE Trades

Equatorial trough

SE Trades

Polar front

▲ Winds blowing north and south from the Equator are deflected eastwards, flung forwards by the speed of the Earth's surface as it rotates at the Equator.

▲ The world wind pattern takes the form of a series of cells in which air rises and falls. The pattern is varied by world geography.

▼ Trees generally grow more vigorously away from the direction of the prevailing wind, giving them a characteristic leaning shape.

Measuring winds

The Beaufort scale gives wind strengths, judged by the effects they produce. Force 1 shows no air movement at all, the strong breeze of force 6 causes sea spray, and force 12 is denoted by the extensive damage caused by a hurricane. The original scale gave wind speeds in miles per hour.

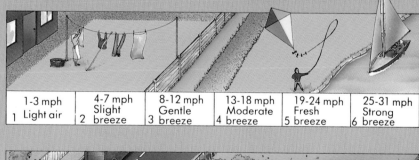

1 Light air 1-3 mph	2 Slight breeze 4-7 mph	3 Gentle breeze 8-12 mph	4 Moderate breeze 13-18 mph	5 Fresh breeze 19-24 mph	6 Strong breeze 25-31 mph

High wind 32-38 mph	8 Gale 39-46 mph	9 Strong gale 47-54 mph	10 Whole gale 55-63 mph	11 Storm 64-72 mph	12 Hurricane 73-82 mph

Measuring the weather

Weather forecasting had always been important to people because of the weather's influence on shipping and navigation, farming and almost every other aspect of human life. Before the development of scientific instruments, weather forecasting relied on observation. People tried to predict the weather by observing such things as wind direction, cloud types, sea colour, and so on. Then in 1643 the Italian physicist Evangelista Torricelli invented the mercury barometer, which measures atmospheric pressure. From that time on, the study of weather became a very much more exact science, now called meteorology.

Modern meteorology contains a number of branches. "Dynamical meteorology" is the branch that deals with the movements of the atmosphere. It takes the sciences of hydrodynamics – the study of movements of liquids and gases – and thermodynamics – the study of the transfer of energy through a system – and applies them to the whole vast ocean of the air.

"Micrometeorology" is the branch of meteorology that deals with more localized effects, such as the development of land and sea breezes, and the formation of valley winds.

▼ Modern meteorologists use radar to help them track the movements of storms. Radar signals are reflected by raindrops and ice particles in clouds.

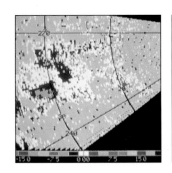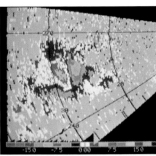

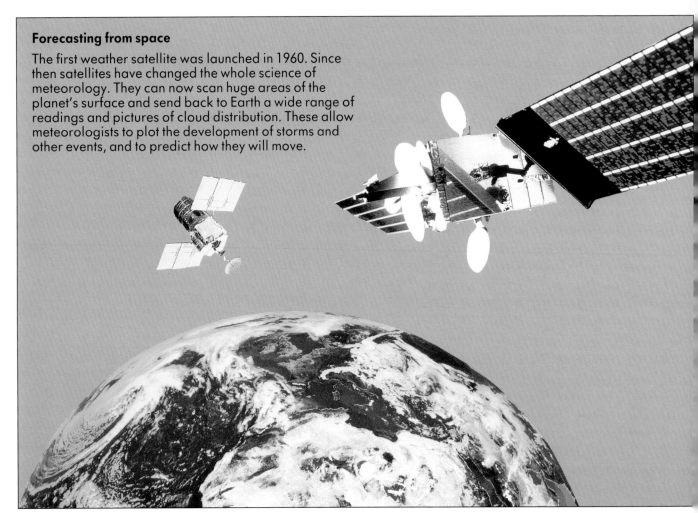

Forecasting from space

The first weather satellite was launched in 1960. Since then satellites have changed the whole science of meteorology. They can now scan huge areas of the planet's surface and send back to Earth a wide range of readings and pictures of cloud distribution. These allow meteorologists to plot the development of storms and other events, and to predict how they will move.

"Synoptic meteorology" is what we normally think of as weather forecasting. Readings of temperature, humidity, air pressure, cloud cover, wind strength and direction, and so on are taken at different places at the same time. This information is then plotted on a chart, or map, to give an overall view of the weather conditions at any particular time. Weather forecasters also use photographs of the clouds taken from orbiting satellites.

▼ A modern weather centre. Information from a large number of weather stations is fed into computers, which produce synoptic charts detailing the weather conditions over a wide area. The charts are used to make weather forecasts.

Processing data

Weather forecasting involves collecting and comparing data from many sources. Much of this is done using a computer system using CPUs (central processing units).

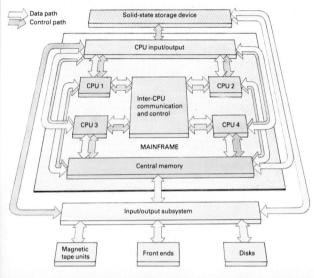

The changing climate

Over millions of years world climates change. The way they have changed can be seen by looking at rocks. In a particular area we might find beds of coal that were produced in a tropical swamp, covered by sandstones that formed in a desert. These may be covered, in turn, by mudstones deposited at the bottom of a shallow sea. Climate changes such as these take place over millions of years.

Extensive changes of climate can take place over shorter periods of time. The ice ages that began 2 million years ago – not a long time in geological terms – did not finish until 10,000 years ago. Throughout that time the world's climate varied widely. At times much of the Northern Hemisphere was choked with ice caps and glaciers. Then a few thousand years later the climates were warmer than they are now. A few thousand years later still the glaciers swept southwards again.

Even in historical times there have been major changes in the climate. On the Tassili Plateau in the middle of the Sahara Desert, there are old rock paintings showing grassland animals. They must have been painted when the local climate was much moister than it is now. Trees still grow nearby. They have immensely long roots which extract water from

▲ The Mexican volcano El Chichon erupted in 1982, sending 16 million tonnes of dust into the atmosphere. The result was an enormous dust veil that absorbed some of the sunlight, leading to a measurable lowering of the Earth's surface temperatures.

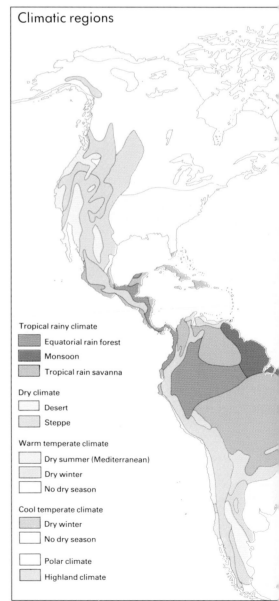

Climatic regions

Tropical rainy climate
- Equatorial rain forest
- Monsoon
- Tropical rain savanna

Dry climate
- Desert
- Steppe

Warm temperate climate
- Dry summer (Mediterranean)
- Dry winter
- No dry season

Cool temperate climate
- Dry winter
- No dry season

- Polar climate
- Highland climate

deep rocks. These trees could not have started growing unless there was water on the surface. Europe suffered a "Little Ice Age" between the 13th and 14th centuries, when climates were very much colder than they are now. In winter, fairs were regularly held on the frozen River Thames, which be impossible nowadays because it does not get cold enough.

The changes in climate through geological time, as revealed by the different rocks, can be explained by the drifting of the continents from one climatic region to another. More recent changes are due to shorter-term events. Volcanic eruptions can throw up dust and gases such as sulphur dioxide high into the atmosphere. There they can block out sunlight and lower the temperatures on the Earth's surface. A noticeable cooling in the 1960s coincided with increasing volcanic activity across the globe.

Another influence may be a fluctuation in the energy output of the Sun itself. Old astronomical records show that the Sun does, indeed, change in size and energy output from time to time. These changes alter the climate.

▼ The Earth can be divided into a number of distinct climatic zones. Changing conditions may mean that maps like this will be inaccurate within a few decades.

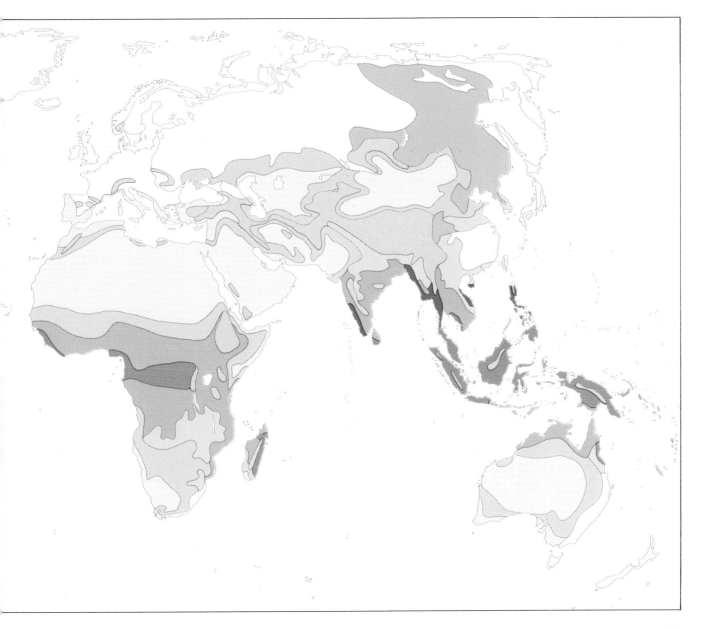

Climates of the world

The Sun's heating effect is stronger at the Equator than at the poles. This produces different surface temperatures over the face of the Earth. It also produces areas of high and low atmospheric pressure, and these generate the pattern of prevailing winds. The distribution of land and sea, the patchwork of the continents and the sweep of the mountain ranges modify the pattern produced. The result is a range of different climates and environmental conditions in different parts of the world. Some are bitterly cold; others are very hot.

▶ Equatorial rain forests flourish because the heat and humidity found along the Equator are ideal for vigorous plant growth. The trees are homes for thousands of different species of animals.

Rain forest climate

The Sun is almost overhead at the Equator. Its rays come vertically downwards, cutting straight through the atmosphere and concentrating their energy on small areas of ground. The air becomes very hot and it rises, producing a belt of low pressure along the Equator. This draws in the Trade Winds from the north-east and south-east. The winds usually travel over oceans and so their air is very moist. When the air reaches the low-pressure areas, it begins to rise. Clouds form and drop their water on the land beneath as daily torrential downpours.

This vigorous circulation of water produces vast networks of streams flowing into the greatest rivers of the world. The Amazon, the Zaire and the Mekong all flow close to the Equator. The humid lowland plains they flow through have a hothouse atmosphere. Plants grow in profusion and produce the tropical rain forest. The conditions are so good for plant life that many thousands of different species can exist in a few square kilometres. They produce trees up to 70 m tall, all growing past each other to reach the sunlight.

Smaller plants called epiphytes grow on the branches, and others in the form of creepers climb up the trunks to reach the light. The intertwined and entangled mass of branches, leaves and creepers forms a green canopy over the whole forest. The tallest trees, the emergents, reach through the canopy into the air above. The forest floor is dark and hot, and few plants grow except where a fallen tree leaves space for the Sun to shine in. Along the river banks the crowns of branches and leaves come right down to ground level.

The vast range of plant types supports a variety of animals as well. Most of these stay among the sunlit branches, although some live in the darkness below.

◀ The gibbon of south-eastern Asia is the most agile of tree-dwelling animals and it eats fruit. The wide variety of plants in the tropical forest has led to the evolution of a range of animals to feed on them.

▼ The several species of sloth from South America are among the slowest-moving of the tree-dwelling mammals. They hang upside-down and live on a diet of leaves.

Grasslands

Grasslands develop naturally in regions where the climate is generally dry but has distinct moist seasons. Grasses can weather long periods of drought because of their underground stems. The leaves and heads may die off in dry weather, but they can grow again from the underground part of the plant. Trees generally do not thrive in such conditions, and the typical landscape is one of wide open plains with very few trees.

The tropical grasslands occur in two bands north and south of the Equator. They lie between the central belt of equatorial rain forest and the two belts of desert along the tropics. As the Earth moves round in its orbit and the Sun appears to move north and south in the sky between summer and winter, the rainy conditions that produce the tropical forest move north and south too. The regions that lie between the forest and desert belts get both types of climate. They have the forest weather at one part of the year, and the desert weather at another. The resulting vegetation is known as tropical grasslands. They can be traced from one continent to another, but they are most prominent in Africa.

▼ The grasslands that we see in moist temperate regions tend not to be natural. Throughout civilization communities have chopped down trees and cut clearings for fields. These take on the appearance of grasslands because the important cereal crops are themselves grasses.

▶ Grassland animals, such as these pronghorn antelopes, have very strong teeth for chewing grass, and complex digestive systems for digesting it. They also have long legs and are built for speed. The best method of defence from predators is to run away.

Temperate grasslands are found deep within continents, usually bordering on cool desert areas. The prairies of North America and the steppes of Asia are the northern examples. In the south, the pampas of South America are partly temperate and partly tropical.

Grassland animals are highly specialized, because grass is a difficult food to digest. The development of grasslands 50 million years ago allowed grazing animals, such as horses, antelopes and cattle, to evolve. In turn, grazing stimulates fresh growth and the animal dung keeps the ground fertilized.

◄ Elephants are not typical of the grassland animals. They are slow-moving, and rather than feed on grass they tend to eat the shoots and leaves of the few bushes and trees that grow there.

▲ The South American rheas, the African ostrich and the Australian emu share the same grassland lifestyle. Flightless birds are typical of grasslands. Their long legs enable them to run quickly over the plains.

◄ The long face of the zebra is typical of a grassland animal. Its mouth can be down munching the grass but its eyes are still quite high and looking about. When zebras graze in herds, there are always one or two animals looking out for danger.

Deciduous and boreal forest

Deciduous trees are those that lose their leaves in winter and grow new ones each spring. The deciduous woodlands are found in the temperate regions of the Earth, mostly in the Northern Hemisphere. The temperate zone is a broad band which, at times, is subjected to the cold wind moving away from the poles. At other times, it is exposed to the warm Westerlies. As a result, the climate is a mild one compared to other climates of the world. It is generally moist and does not have extremes of heat or cold.

This zone has particularly favourable places for people to live and grow crops. Over the centuries, much of the original deciduous forest has cleared away for cities and farms. The common large trees of these woodlands are broad-leaved types, such as oaks, ashes, beeches and willows. Smaller trees growing beneath them include maples and birches. At a lower level still grow the bushes of dogwoods, hollies and hawthorns, and there is usually a thick undergrowth of flowering plants.

To the north of the belt of deciduous woodland lies the largest stretch of uninterrupted forest in the world. The great coniferous boreal forest stretches from Scandinavia eastwards across northern Europe and Asia, then across Alaska and Canada. There the prevailing weather is brought by the cold air masses that blow in from the far north.

The growing season is only three or four months long, and during the lengthy winter all the moisture is locked in ice and snow. The coniferous trees can withstand these conditions.

▶ Deer are the typical animals of deciduous woodland. They eat a number of different foods, including leaves and young shoots from the trees and the undergrowth.

▼ A deciduous woodland has several different kinds trees, with a spread of bushes and an undergrowth of many small flowering plants.

The needle-shaped leaves of coniferous trees reduce the rate at which water is given off through them. The leaves stay on the tree all year round, so that they are ready as soon as the growing season begins. The trees' conical shape allows snow to slide off easily.

Although coniferous trees are particularly well adapted for the cold conditions, they are found in more temperate regions as well. The deciduous woodlands rarely consist of only deciduous trees but usually have conifers among them – giving a "mixed woodland".

▲ There are fewer animals and birds in coniferous forests than in deciduous woodlands, and they tend to be quite specialized feeders. Crossbills (1), for example, eat only seed cones. The coniferous forests of Canada (2) are home to the beaver – a creature that can alter its habitat. Beaver colonies can fell trees and dam rivers, holding back lakes in which they build their lodges. The blunt face of the beaver (3) hides a massive set of chisel-like incisor teeth. It uses these to fell the trees, and groups of about 12 cooperate in building the dams and lodges.

Mountain climate

Conditions change as we climb a high mountain. The higher we go, the thinner the air becomes. The sky also gets bluer and the Sun's rays become hotter. There is a change in climate between the mountain's base and its summit. The difference can be similar to the difference between the climate at the Equator and at the North Pole or South Pole.

At the base of an equatorial mountain, such as Mount Kenya or Mount Kilimanjaro in eastern Africa, there may be tropical forest or tropical grassland. Higher up the temperature falls, and the climate becomes moister because of rain falling from the rising clouds. At about 1,500 m the original tropical conditions give way to those of moist temperate forests.

Above about 2,400 m there is less rainfall and the forest gives way to scrub. Grasses become the main plants. In the African and Asian mountains, bamboo is the most common.

The bamboo and scrub give way at a height of about 3,000 m to open moorland, with tussocks of coarse grasses and heathers. Up there the conditions are too harsh for trees to grow, and the landscape is similar to that of the tundra regions of the far north.

Above the tree-line

Just as the tundra in the north gives way to the ice-caps of the North Pole, so the moorland of the mountain gives way to the glaciers and snow-covered crags of the mountain summit. On top of the African mountains there are permanent snowfields and glaciers, even though they lie near the Equator.

Most mountain slopes are gentle, especially among the foothills. Changes in climate usually take place over quite large distances. But in the Himalayas, the gorge of the River Brahmaputra is so narrow and steep that conditions change from tropical forest to snow and ice within a kilometre or two.

▶ Big horn sheep in Alaska. The change in climate between the base of a mountain and its summit is accompanied by a change in plant and animal life. Typical forest animals such as pigs and deer give way to more specialized creatures on the higher slopes. In Asia, giant pandas live at this level, and in Africa there are mountain gorillas. Sure-footed mountain goats and ibexes live on the sparse grazing of the higher reaches.

Ice

Tundra

Scrub

Cloud forest

Rainforest

Arid ground

Deserts and polar climates

Plants need a certain minimum amount of water to survive. Very few plants can grow in particularly dry regions, where there is very little rainfall or all the moisture is locked up as ice. The barren landscape that results is known as a desert.

Hot deserts are the kind that usually come to mind when we think of a desert. There are several types of hot deserts. Tropical deserts lie in two belts along the Tropics of Cancer and Capricorn. Hot air that has risen and released its moisture as rain over the low-pressure equatorial belt now descends over the tropics. This dry air forms high-pressure belts and no moist winds blow over these areas. The great deserts of the Sahara and Arabia lie in this zone in the north. In the south there are the Kalahari in Africa and the Gibson in Australia.

Continental deserts exist in places that are so far from the sea that the moist winds just cannot reach them. The Gobi Desert in central Asia is a typical example.

Finally there is the rain shadow desert, which is found on the lee side of mountain ranges (the side away from the wind). Winds from the ocean lose all their moisture in rain as they rise up the seaward mountain slopes, leaving only dry air to pass over to the other side. California's Death Valley is the most famous rain shadow desert.

Hot deserts are usually surrounded by zones of semidesert, in which only specialized types of plants can grow. This is also true of the cold deserts. Beyond the northern reaches of the great coniferous forests there is a region in which the climate is too harsh for trees. It is an area of permanently frozen subsoil with a vegetation of coarse grasses, heathers and

▲ Cactus plants have a thick fleshy stem that holds water, a leathery waterproof skin to keep the moisture in, and spines to protect them from damage by animals. All these features enable them to survive in harsh desert conditions.

◄ Death Valley in California is the hottest place on Earth. For most of the year it is totally dry. But when the rains do come they occur as torrential downpours with several centimetres falling in a day. Immediately afterwards the desert blooms with all kinds of plants before it returns to barrenness for the remainder of the year.

other small herbaceous plants. It is called the tundra where it occurs in Europe and Asia, or the muskeg in Canada.

Islands close to the North Pole and the continent of Antarctica at the South Pole can be regarded as true deserts. As a result of the Earth's tilt, the Sun shines there for only part of the year, and when it does its rays slant through thick layers of atmosphere. Because of the permanently low temperatures water is always frozen and useless to living things.

▼ The icy lands of the far north are barren enough to be considered as deserts. However, the sea supports a great deal of life. Algae grow in the waters, and are eaten by fish. The fish are food for seals which are, in turn, hunted by polar bears. Birds, such as gulls, act as scavengers.

▲ Small cushion-like plants grow all over the tundra. Their compact shape gives them protection against the severe weather.

▶ Reindeer graze on the tundra vegetation in summer, but migrate southwards to the forest in winter.

Origins of life

Spot facts

• *Four of the elements essential to life (carbon, hydrogen, oxygen and nitrogen) are among the five most abundant elements in the Universe.*

• *There is just a faint possibility that life might exist elsewhere in the Solar System. Traces of amino acids – the building blocks of life – have been found on some meteorites.*

• *The earliest evidence of life on Earth is a fossil bacterium 3,200 million years old.*

• *From a few fragments of fossil bones and teeth, scientists can work out what a prehistoric animal looked like and what it probably ate.*

▶ Chemical reactions in the early atmosphere were fuelled by the energy of lightning, ultraviolet radiation from the Sun and the heat of volcanoes. The chemicals produced then formed a thin "soup" in the oceans, where the earliest life began.

Our planet was formed about 4,600 million years ago. Some time during the first 1,000 million years, life in its most basic form appeared on Earth. There is a theory that the Earth might have been "infected" with life from outer space. But life was probably created from simple chemicals that occurred in the early Earth's atmosphere and in the oceans. Gradually, natural chemistry produced more complex substances that became the building blocks of life. For the next 3,000 million years, life was confined to the seas, where it began. Traces of early lifeforms can be found as fossils embedded in rock.

Fossils

Fossils are the remains of animals and plants that over millions of years have been converted into stone. They are evidence of early life on Earth, and are usually found embedded in other rocks. Fossils range from faint traces and footprints, to complete plants and animals, showing details of size and shape.

When a living thing dies, it does not always decay. Sometimes it is covered with layers of mud, which gradually turn into rock. The structure of the plant or animal also turns to stone and is preserved as a fossil. Much of what we know about the evolution of life and prehistoric plants and animals comes from the study of fossils.

The oldest fossils

In the 1950s fossil hunters began finding evidence of the first living things. Microscopic fossil bacteria have been found in rocks dating from more than 3,000 million years ago. The oldest fossils visible to the naked eye are about 1,800 million years old. They are the remains of stromatolites, structures built up by colonies of millions of single-celled plants.

▲ A cross-section through a fossil stromatolite shows how millions of single-celled plants were organized in layers to form a colony.

▼ Structures similar to fossil stromatolites are still being formed in warm, shallow seawater. For example, they can be seen as large lumps, covered in seaweed, at low tide on the coast of South Africa.

Dawn of life

For 1,000 million years, chemical reactions in the world's ancient oceans produced a thin "soup" of complex organic chemicals. As the soup simmered on hot rocks at the edges of the seas, cells developed that were able to reproduce themselves. These earliest cells are known as prokaryotes.

About 2,500 million years ago, one group of prokaryotes became able to use the energy of sunlight to make food. This was the beginning of photosynthesis. These cells eventually became the first primitive green plants. Prokaryotes still exist today as bacteria and the closely related blue-green algae.

The first complex cells
Early simple cells gradually developed into more complex ones. Prokaryotes started forming communities of cells, such as those forming

▶ Life in the early oceans. Jellyfish (1) and burrowing worms (5) look much the same as their present-day descendants. The two species that are anchored to the sea bed (2 and 4) resemble modern sea-pens, a form of soft coral. The circular worms (6) and the strange disc-shaped creature (3) are unlike any known animals. They must have either died out or evolved into something else. None was very big. The worms were about 2 cm long, and the sea-pens may have grown 12 cm tall.

Invertebrate fossils

1 Cyclops medusa	4 Charnodiscus
2 Glaessnerina	5 A burrowing worm
3 Dickinsonia	6 Tribrachidium

stromatolites. But the great leap forward came about 1,200 million years ago, with the appearance of cells with a nucleus. The nucleus is a central core that controls the cell's activities. These more complex cells are called eukaryotes. They gave rise to the great diversity of life on Earth. The first eukaryotes were probably simple plants and tiny single-celled animals called protozoa.

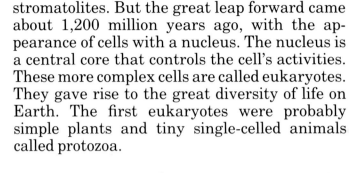

◀ The earliest animals have left only faint traces. The fossil worm tracks (left) are clearly identifiable, but tell us practically nothing about the worm. The "mould" of an external surface (right) shows that the animal was made up of segments.

Multi-celled life

The first animals and plants that consisted of many cells appeared about 1,000 million years ago. The first animals were small and had few hard parts. For this reason, they did not readily form fossils and so they have left very little trace of their existence.

Some of the earliest fossils are the tracks and tunnels made by soft-bodied worms at the bottom of the sea. All the available evidence strongly suggests that by about 680 million years ago, there were already many different forms of life, concentrated in shallow seas. They were poised for the next evolutionary step, the development of hard skins and shells that provided outer protection.

▲ A fossil sea-pen. The animal resembled a feather, growing out of the seabed. Not until animals developed hard shells do fossils become more informative.

Record in the rocks

Fossils in the rocks tell the story of the development of life on Earth. We can read the story in the correct order because the rocks started as layer upon layer of mud and other sediments on the bottom of ancient seas. Each layer took millions of years to form, and they are often thousands of metres thick and very deep down. Only in a very few places are rocks older than 600 million years near enough to the surface for us to find fossils in them.

The study of fossils is called palaeontology. This science first began as a branch of geology, the study of the Earth. Then biologists began to become interested in fossils to date stages of evolution. If a fossil is found in a rock of a certain age, that must also be the age of the fossil.

Imperfect record

The fossil record is far from perfect, and new discoveries are being made every year. Fossils are found only when rocks are exposed at the surface, or during mining and tunnelling. The folding and movement of the rock layers, and the action of rivers and earthquakes, have exposed areas of rock from all periods of the Earth's history.

Some rocks provide better fossils than others. Rocks are made up of tiny particles called grains. Fine-grained rocks, such as slate and limestone, provide much better quality fossils than coarse-grained rocks, such as sandstone. From fairly recent times we have some almost perfect specimens that have been preserved in drops of amber (fossil resin from trees) and natural deposits of tar.

Fossils provide a wide range of information. The discovery of a number of different species in the same small area of rock helps us to form a picture of the community of life at that time. Fossils can be used to trace how a species has developed over hundreds of millions of years. The fossil record also has remains of species that no longer exist. Scientists use such fossils to work out what the creatures looked like.

◀ This near-perfect fossil skull of a sabre-toothed tiger is about 2 million years old. Scientists can take accurate measurements from such well-preserved specimens and compare them with modern species.

▼ Ammonites are one of the most common fossils, and there are hundreds of different varieties. They flourished in the oceans for more than 300 million years, then suddenly vanished from the fossil record.

Cross-section of rocks

The Grand Canyon provides spectacular scenery, and a slice through geological time. The Colorado River has slowly cut down through 1,500 million years of rock layers. The present topmost layer consists of Permian rock about 250 million years old. A little way down the Canyon side are fossilized sea creatures that lived 300 million years ago. Lower down, there are fossil shellfish from 500 million years ago. At the bottom are rocks dating from about 1,700 million years ago.

Geological timescale

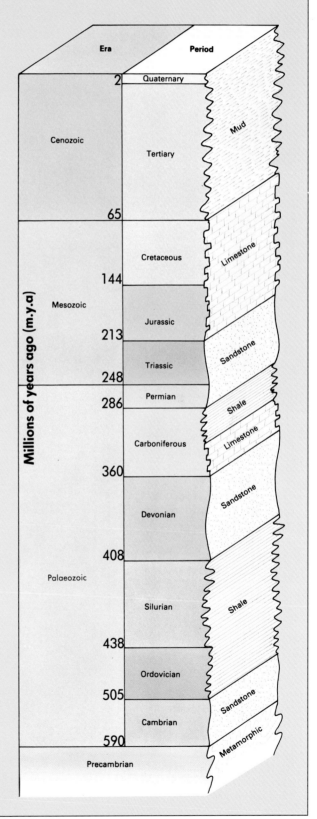

Era	Period	
		2
Cenozoic	Quaternary	
	Tertiary	Mud
		65
Mesozoic	Cretaceous	Limestone
		144
	Jurassic	
		213
	Triassic	Sandstone
		248
	Permian	286
		Shale
Palaeozoic	Carboniferous	Limestone
		360
	Devonian	Sandstone
		408
	Silurian	Shale
		438
	Ordovician	
		505
	Cambrian	Sandstone
		590
Precambrian		Metamorphic

Millions of years ago (m.y.a)

Explosion of life

Spot facts

● *The first meat-eating animals were probably ancestors of the squid.*

● *The first animals to breathe air and walk on dry land were scorpions.*

● *During the "Age of Dinosaurs" no other kind of land animal bigger than a large hen walked on the Earth.*

● *Some dinosaurs weighed more than 60 tonnes when fully grown; others probably weighed only about 10 kg.*

● *The death of the dinosaurs made room for giant mammals, such as a rhinoceros 6 m tall.*

▶ A prehistoric footprint. Some of the most dramatic fossils ever found are dinosaur footprints, formed in mud and then preserved as rock for more than 80 million years.

Once animal life became established in the seas, it developed very rapidly. Different animal groups pioneered each new development: jointed legs, eyes, claws and backbones. About 395 million years ago came a great breakthrough: life moved on to the land. By this time many animals had reached their present-day forms. But life kept on developing. The first four-legged animals, the amphibians, gave rise to a new group – the reptiles. Today, there are still a few amphibian species, such as frogs, and various reptiles, like snakes and lizards. But the most highly evolved animals are the mammals.

The Cambrian

The Cambrian Period (590-505 million years ago) saw the establishment of early forms of most present-day invertebrates, the animals that have no backbones. Animals resembling sponges and worms lived on the seabed. But the pace of development was set by the trilobites. These were among the first arthropods, animals with jointed limbs, like crabs and insects. They were also the first animals to develop really efficient eyes.

Arthropods are thought to have evolved from worms which developed a pair of legs on each of the many segments that made up their bodies. The earliest ones still look like worms. The trilobites looked rather like woodlice, although there were thousands of different species. They were highly mobile herbivores (plant-eaters),

and they dominated the Cambrian seas. The trilobite "design" was very successful, but it had its limitations. Other arthropods developed specialized limbs that could be used for grasping and grabbing. This enabled them to tackle different foods.

The first meat-eaters

The other widespread animals in the Cambrian seas were the brachiopods. They were a kind of armoured worm, resembling a present-day shellfish. Towards the end of the period there was a very noticeable increase in the number of fleshy animals with hard shells. The shells acted as protective armour. This was a necessary defence against the first carnivores, or meat-eating animals.

Key
1 Brachiopod
2 Trilobites
3,4,5 Arthropods
6 Sea lily (crinoid)
7 Sponge
8 Sea anemones

▼ Trilobites and brachiopods make up 90 per cent of Cambrian fossils, but many other lifeforms flourished. All the animals shown here shared the same habitat about 550 million years ago. The trilobites display their characteristic shape, and the smaller arthropods are developing primitive claws. The sponge, sea anemone, and sea lily have already acquired their basic present-day shape.

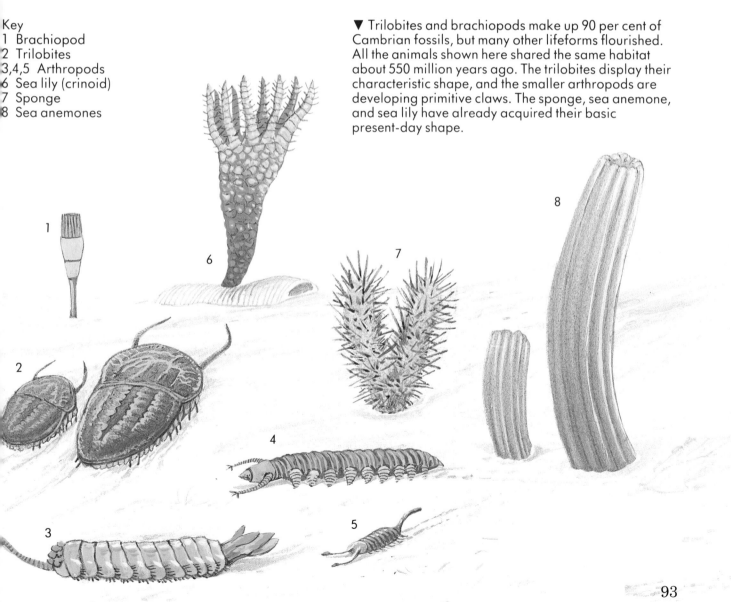

The Ordovician

During the Ordovician Period (505-438 million years ago), invertebrates continued to develop. Many gained their present-day characteristics. The headless animals, such as starfish and sea urchins, began to populate the seabed in large numbers. And sponges, corals, sea cucumbers and molluscs all developed their basic shapes and flourished. Towards the end of the period, the first sea-scorpions emerged. Their well-developed pincers, positioned in front of the mouth, made them much more efficient than the trilobites at gathering food. They became underwater hunters.

Into deeper water

Animal life slowly began to extend out of the shallowest water, as first brown and then red algae (seaweed) moved progressively deeper. Around the beginning of the period, tiny, half-worm, half-fish creatures called lancelets appeared. They had a kind of spine called a notochord. The notochord marks the beginning of the development of the spinal chord and the backbone. By the middle of the Ordovician Period, the first primitive fish had evolved. These were small creatures 5-8 cm long, which fed by sucking in water and filtering out tiny animals and plants. They are generally considered to be the first vertebrates (animals with a backbone). The spine of such fish was very weak; for support they had instead a tough skin consisting of interlocking plates, which acted like a suit of armour.

▼ A fossil trilobite. Trilobites flourished for more than 200 million years and were one of the most successful groups of animals of all time. They were arthropods that lived on the seabed. A few were swimmers, and some burrowed into the mud. The spines gave an increased area without adding much weight. Trilobites became extinct about 300 million years ago.

The Silurian

Salt and freshwater fish

The Silurian Period (438-408 million years ago) was a time when life swarmed in the shallow seas. Corals built up huge reefs, providing a place for many types of invertebrates to live. During a relatively short span of time (45 million years) two major animal types – the fish and the arthropods – made great progress in their development.

Jawless fish, which fed by sucking in their food, gradually invaded freshwater rivers and lakes. In these quiet inland backwaters, the first fish with jaws evolved. These were also the first that fed on other fish, and they remained confined to fresh water. A little later in the sea, fish called placoderms appeared. They were covered in hard, bony plates. Ranging in size from a few centimetres up to 9 m in length, they were powerful predators.

Scorpions on the move

On the sea bottom, sea-scorpions were evolving rapidly. The development of grasping claws gave them a strong advantage as predators. Some also developed strong paddles for swimming. One species was more than 2 m long.

◄ ▼ The fossil shellfish on the left has been magnified 700 times to show its intricate internal structure. Countless billions of such tiny animals lived in the ancient seas. Over millions of years their minute skeletons formed thick layers of sediment on the seabed which gradually turned into rock. The cliffs in the picture below are made of such rocks.

▼ Layers of sediment on the sea bottom form bands of rock called strata, such as these layers of chalk and clay. (The coin is there to show scale). Under a microscope we can see thousands of tiny fossils in such rock.

The Devonian

At the very end of the Silurian Period (about 408 million years ago) both plants and animals first made the great transition from water to land. Some of the first land plants with a system for transporting water inside their structure had stems, but neither roots nor leaves. The first air-breathing animals were almost certainly small scorpions. Their legs and claws gave them the mobility that was needed to catch prey in the new environment.

The Devonian Period (408-360 million years ago) is sometimes called the Age of Fishes. The new kinds of fish included the sharks, with their cartilage skeletons, and the bony fishes. All had very strong backbones. By the end of the period they had taken over from the earlier types. There were two kinds of fish with bony skeletons. One kind (called ray-finned) had ordinary fins, and the other (bony-finned) had bones that extended into their fins.

Many forms of sea-life have been excellently preserved as fossils. The strong shells of the brachiopods (1) are easily identifiable, as is the typical shape of the sea lily (2). The long tapering internal skeleton of the squid (3) is more of a puzzle, because the soft body shape and tentacles have not been preserved. On the fossil of a fish (4) even individual scales can be seen.

All the descendants of the sharks and ray-finned fish have remained in the sea. But some of the bony-finned fish, which lived in the shallowest water, began to develop the ability to breathe air. At the same time, their fins became more muscular, enabling them to scramble about on land. These fish are now extinct. But during the Devonian Period, some of them evolved into new, four-legged creatures that breathed air. These were the amphibians.

Land plants had by now developed into large tubular structures that were efficient at sucking moisture from the soil. The first swampy forests appeared, full of giant horsetails and ferns. These forests were home to the first wingless insects, the springtails and bristletails. The first millipedes and spiders also appeared. By the middle of the period, the arthropods had established themselves as the dominant land animals.

Late Devonian lakeside

Amphibians were the first vertebrates to walk on dry land. They were also the first four-legged animals. *Icthyostega* (below) was one of the first true amphibians, although fossils reveal that its tail was still rather fish-like. The legs, which had evolved from the fish's bony fins, were well developed. This shows that it probably spent most of its time on land. The adults were about 1 m long, and probably fed on insects. On land they had nothing to fear except the scorpions, which by this time had developed a powerful sting in their tails. Like all amphibians, *Icthyostega* had to return to the water in order to lay its soft-shelled eggs. These were probably rather like frogspawn, and hatched into tadpole-like creatures.

The Carboniferous

Huge swampy forests and jungles covered the Earth during the Carboniferous Period (360-286 million years ago, or m.y.a.). The ground lay thick with rotting vegetation that was later to become coal. Insects had taken to the air by this time and many different species of cockroaches and centipedes had also appeared. But above all this was the Age of Amphibians, which were ideally suited to the damp conditions. Those that spent most of their time out of the water had strong, muscular legs. Some of the later species grew as big as alligators.

The Permian Period (286-248 m.y.a.)

During this period, some of the amphibians developed the ability to lay hard-skinned eggs, which could survive on land without drying out. These were the first reptiles, but only just. Their fossils are often almost identical to true amphibian fossils. By the end of the period many different reptiles had appeared, and had colonized the dry land away from water. Then suddenly about half of all animal species became extinct. Sea-scorpions and trilobites disappeared from the seas. Most large amphibians and many of the early reptiles also died out. These extinctions mark the end of the Palaeozoic Era, the time of primitive life.

The Triassic Period (248-213 m.y.a.)

A new group of animals, the dinosaurs, emerged from the disaster. At first quite small, the dinosaurs rapidly developed a wide variety of shapes and sizes during this period. The first lizards, tortoises and crocodiles also appeared around this time. Another group of reptiles developed some of the characteristics of mammals. Some were warm-blooded and suckled their young. The first true mammals, tiny insect-eaters the size of shrews, also appear for the first time as Triassic fossils.

▶ The forests of the Carboniferous Period were dominated by huge primitive conifers (1), club mosses (2, 6) and horsetail creepers (4) climbing up around the tree-trunks. The undergrowth was a tangle of tree ferns (3), horsetails (10) and club-moss roots (5). Dragonflies (7) flitted through the damp air, and the vegetation provided food for centipedes (9). Many amphibians (11) grew quite large. Those that stayed in the water (8) had a much more eel-like appearance.

The Jurassic

The Jurassic Period (213-144 million years ago) was the height of the Mesozoic Era, the time of middle life. The mammal-like reptiles soon became extinct and the dinosaurs and true reptiles took over. Some dinosaurs were the largest land animals that ever lived, and some were no bigger than a large hen. For 100 million years, on land, in the air and in the water, these animals ruled the Earth. Herds of large and small herbivores, some armoured, some not, grazed the vegetation. The largest dinosaurs weighed 80 tonnes and stood over 20 m tall. They evolved long necks in order to reach the highest leaves.

Other species developed into ferocious carnivores with sharp fangs and teeth. Huge flying lizards preyed on their smaller relatives, some of which were gradually evolving into primitive birds. At sea, the icthyosaurs (fish-lizards) bore a remarkable resemblance to the modern dolphin. The insects also increased in variety at this time. Bees, wasps, ants and flies all evolved during the Jurassic Period. Fossils show that mammals were still very small.

The Cretaceous Period (144-65 m.y.a.)

During this period, the dinosaurs reached the height of their evolution. Many had become fairly lightweight, fast-running species. Others had evolved into tank-like armoured giants, or ferocious flesh-eaters like *Tyrannosaurus*. But other life was also evolving. Snakes, birds, flowering plants and trees all emerged in their present forms at around this time. Mammals were also developing noticeably, using different methods of reproducing instead of laying eggs. Then suddenly there was another catastrophe of some kind, and dinosaurs died out. The death of the dinosaurs marks the end of the Mesozoic Era, and the beginning of modern life.

▶ *Rhamphorhynchus* (1) soared high above the Jurassic landscape. On the ground, herbivores fed on the foliage of conifers and ginkgoes (2). The largest, such as *Brachiosaurus* (3), had little to fear. Their smaller relatives, such as *Dicraeosaurus* (4), fled into shallow water at the approach of a predator like *Megalosaurus* (5), only to be threatened by crocodiles (9). Stegosaurs (6) were protected by spiky armour while they grazed on ferns and horsetails (10,11). Smaller dinosaurs (7, 8) relied on speed to escape.

The Tertiary

The Tertiary Period (65-2 million years ago) was the beginning of the Cenozoic Era, the time of modern life. It was marked by a tremendous increase in the different kinds of mammals.

Nearly all modern mammal species started as small rabbit-sized creatures. Rodents, such as rats, appeared about 50 million years ago. Bats, primates and whales developed about 10 million years later. By about 30 million years ago, cats, dogs, horses and pigs had joined the growing ranks of mammals.

Between 26 and 27 million years ago, mammal life reached its greatest ever diversity. There were thousands of variations on present-day types. The largest land mammal ever, a giant rhinoceros more than 6 m tall, lived about 20 million years ago.

During the later part of the Tertiary Period, the great continents finally achieved their present shapes. But because of lower sea levels, there were land bridges between continents – for instance, between Asia and North America. Mammals that had evolved in one part of the world could now begin a series of great migrations. Elephants spread from Africa to America and Eurasia; pigs and cats moved in the opposite direction. By the end of the Tertiary Period, all life on Earth had evolved its basic present-day forms.

The Quaternary Period (2 m.y.a. – today)

During the most recent geological period, many other mammals grew to enormous size compared with their modern descendants. There were giant deer with antlers 3 m across. During this period there were several ice ages, when temperatures were much colder than normal. By the end of the last ice age, about 10,000 years ago, all the giant mammals – except whales – had died out. Life on Earth was much the same as we see it today.

▶ By the last part of the Tertiary Period, deciduous trees and flowering plants had spread throughout the Earth. In this African scene, ancestors of the elephant *Amebeledon* (1) and camel *Procamelus* (2), share a waterhole. In the background ancestors of the horse gallop across the plains. *Neohipparion* (3) had three toes, and had not yet developed proper hooves. *Pliohippus* (4) was the first of the one-toed horses.

The human ape

- *The oldest human fossils are of a young woman who has been named Lucy; she lived over 3 million years ago.*

- *The first industries developed about 1½ million years ago, when primitive people began making stone axes.*

- *The "cave-men" used caves for shelter, but most people probably lived in tents or wooden shelters.*

- *Neanderthal people were far from stupid. Their brains were actually bigger than the modern human brain. Nobody knows why the Neanderthals disappeared about 35,000 years ago.*

▶ These footprints clearly preserved in stone show that two adults and a child once walked across a muddy area of Africa about 4 million years ago. The way the footprints are placed tells us that by this time our distant ancestors walked upright.

The story of humans is both fascinating and incomplete. Many parts of the story are known, but there is a large gap in the fossil record during a critical stage of our development. Our closest living relatives are the apes. However, we are a very special kind of ape. We have learnt how to speak, and how to make and use tools. Since humans first walked out of Africa about 1 million years ago, we have spread to every part of the world, and adapted to many of its environments.

104

Our ancestors

▼ Skulls are the most durable of hominid remains. They enable us to measure the size of the brain. We can trace our ancestry through the development of the skull.

Family tree

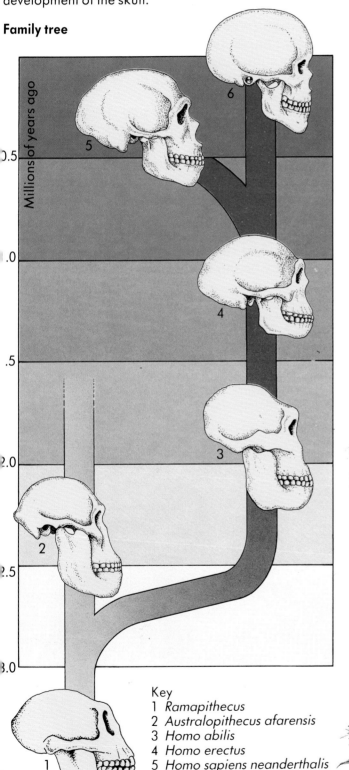

Millions of years ago

Key
1 *Ramapithecus*
2 *Australopithecus afarensis*
3 *Homo abilis*
4 *Homo erectus*
5 *Homo sapiens neanderthalis* (Neanderthal people)
6 *Homo sapiens sapiens* (modern people)

Fossils tell us that between 14 and 8 million years ago a type of ape called *Ramapithecus* was widespread from present-day France to northern India. The first *Ramapithecus* fossil was discovered in northern India in 1932.

Ramapithecus was probably the first relative of the hominids, human-like apes. It probably lived in partially wooded regions and spent a lot of its time on the ground. It may have left the trees altogether. All of its weight would certainly have been on its back legs while feeding. This would have freed its hands to gather more food such as roots, nuts and seeds. When walking any distance, however, *Ramapithecus* almost certainly used all four limbs, rather like a modern chimpanzee. No hominid fossils have been found dating from the 4-million-year period following the disappearance of *Ramapithecus*.

▼ *Ramapithecus* was very ape-like in appearance. Collecting food on the ground, and not in the trees, probably encouraged the habit of walking upright.

Early people

About 4 million years ago, parts of what is now eastern Africa were inhabited by small hominids who walked upright all the time. These are our earliest direct ancestors. We have no fossil remains of them, but we know that they developed into two subgroups. One of these groups is called *Australopithecus*. The earliest fossil hominid is of a female, whose bones are 3.2 million years old. She was discovered in Ethiopia by an American scientist, John Johanson, in 1974, and named Lucy.

By about 2.5 million years ago, two distinct kinds of *Australopithecus* had emerged. One was larger and stronger, and is called the robust type. The other was more slender. It is called the gracile type. Both probably carried sticks and threw stones. But no tools have ever been found with their remains. All types of *Australopithecus* became extinct about 1 million years ago.

Nothing is known about the other side of Lucy's family tree until around 2.5 million years ago, when the first true human appeared. The earliest type is known as *Homo habilis* "skilful man". The earliest fossil remains of *Homo habilis* coincide with the first real tools – crudely shaped pieces of stone. Big stones were used for smashing up animal bones to get at the marrow, and sharp flakes of rock were used to cut up the meat.

About 1.5 million years ago the next type of human emerged, *Homo erectus* "erect man". They were able to make better tools, by chipping stones into cutting tools called hand-axes. Groups of *Homo erectus* in different areas of Africa developed different techniques of making stone tools. By about 1 million years ago, *Homo erectus* had learned to make use of fire, and had invented cooking. At about that time, they began to spread out from Africa.

H. afarensis

▼ Lucy belonged to the earliest type of *Australopithecus*, which is known as *afarensis*. Her brain was only about a quarter the size of a modern human's. Although her descendants grew taller and stronger, their brains remained about the same size and they never learned to make tools.

▶ The brain of *Homo habilis* was about half the size of that of modern people but was twice as large as that of *Australopithecus*. This increase in brain power enabled *Homo habilis* to learn how to make the first crude stone tools, which could be used to cut up dead animals.

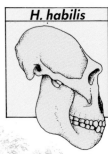

H. habilis

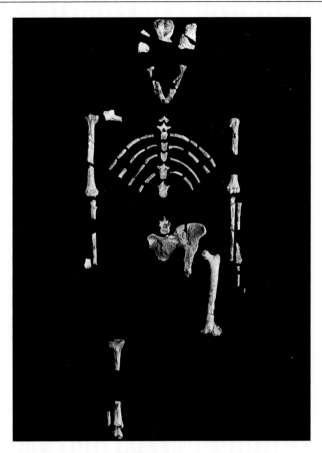

An apewoman called Lucy

Lucy is the nickname of the fossil of a female *Australopithecus*. By studying the knee and hip joints scientists are certain it was a female who walked upright like a modern human. She was 1.2 m tall and died at about 23 years old.

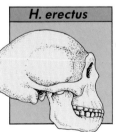

H. erectus

Homo erectus had a brain about two-thirds modern size. The increase in brain size over *Homo habilis* was quite small. However, it was enough to enable *Homo erectus* to develop quite sophisticated stone tools, and to learn how to make and use fire. Once *Homo erectus* had fire, they moved north to cooler climates, using the discovery to keep warm.

Prehistoric people

We know from fossils that *Homo erectus* had spread into Europe and the Far East by around 500,000 years ago. By this time these early humans had learned to make semi-permanent campsites where the hunting was good. The examination of an early European campsite (made about 400,000 years ago) has shown that rhinoceroses were the most common prey. Presumably, this was because they provided plenty of meat. Beavers were also hunted in large numbers for their warm fur.

Over the next 200,000 years, *Homo erectus* slowly evolved into *Homo sapiens* "wise man". By about 200,000 years ago, the transformation from one group to the other was complete, but human evolution was not quite finished. Two further subgroups developed: *Homo sapiens sapiens* "modern man" and Neanderthal people.

These were the so-called "cave-men", but the term is misleading. Both groups used caves to shelter from the cold of the ice age and from dangerous animals, such as the sabre-toothed tiger. But both groups also made tents and wooden shelters, and inhabited grassland and forests. The reason these prehistoric people are associated with caves is that the best-preserved evidence of human occupation has been found in caves. In the same way, stone tools are the only ones to survive. But prehistoric people undoubtedly made baskets and nets, which have long since crumbled into dust.

Neanderthal people first appeared in Europe about 120,000 years ago. They are often said to be stupid and beast-like, but nothing is further from the truth. The Neanderthals were skilled stoneworkers, and are the first people known to have buried their dead. In one grave that was discovered recently, a bunch of flowers had been placed with the body.

Homo sapiens sapiens first appeared in southern Africa about 100,000 years ago, and these people quickly developed the skills that would make them the masters of the planet. Their stone tools became increasingly more specialized and finely crafted. By the time they moved northwards about 40,000 years ago, they had learned how to carve bone and ivory into spearheads, needles and combs.

Neanderthals

▼ The brain of the Neanderthals was larger than that of modern people. Around 35,000 years ago Neanderthals died out and we do not know why. Some scientists have suggested they were killed off by *Homo sapiens sapiens*, but this seems unlikely.

A painting of a bison on the roof of the cave of Altamira, in the Pyrenees Mountains in Spain. It is a superb example of prehistoric art. The earliest European cave paintings date from around 20,000 years ago, and depict the animals that were hunted by prehistoric people. They may have been painted for religious or magical reasons.

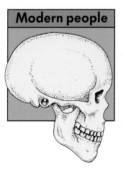

Modern people

Tents made of animal skins were a much more practical form of shelter than caves. They could be packed up and carried if the animal herds that were hunted for food moved away.

▼ These prehistoric implements range from crudely shaped hand-axes to finely formed arrowheads.

The human race

Human migration from southern Africa began at least 50,000 years ago. Towards the end of the last ice age, about 20,000 years ago, *Homo sapiens sapiens* was widely spread throughout the warm moist regions of Africa, Europe and Asia. They did not move into North America until about 15,000 years ago, when the first settlers crossed over the frozen sea between Siberia and Alaska. They soon spread southwards, and reached the southern tip of South America by about 12,000 years ago.

People have continued to migrate in more recent times. Successive waves of people moved eastwards into Europe between 500 BC and 1000 AD. And during the last two centuries, millions of people have travelled to make a new home in the Americas.

Different races

To a biologist, a race is a large subgroup of a species that lives in a particular region and differs slightly in appearance from other subgroups. Humankind is usually described as being divided into five separate races. Nobody knows when these divisions first began.

We are all *Homo sapiens sapiens*, and there is no evidence of any variation in intelligence between the human races. But there are enormous variations in height, skin colour, and facial structure. For instance, dark skin is thought to provide better protection against strong sunlight. Each race has its own group of languages. More than 99.9 per cent of the world's population consists of people who are Caucasoid, Mongoloid or Negroid or a mixture of more than one race. Only a few people remain of the Australoid and Khoisan races.

▶ **Australoid** people have dark skins and flat noses that are suited to the hot, dry conditions of central Australia. **Caucasoid** people show the greatest variation in skin and hair colour, from the blonde-haired and fair-skinned people of northern Europe, to the dark-haired and dark-skinned people of India. **Khoisan** people, from southern Africa, have yellowish skin and their own distinctive language. **Mongoloid** people have the least body hair. Their skin colour varies. The Chinese have pale skins, the American Indians, dark skins. **Negroid** people have dark skins, and display the greatest variation in height. Both the tallest and shortest peoples of the world belong to this racial group.

The human race

Negroid

110

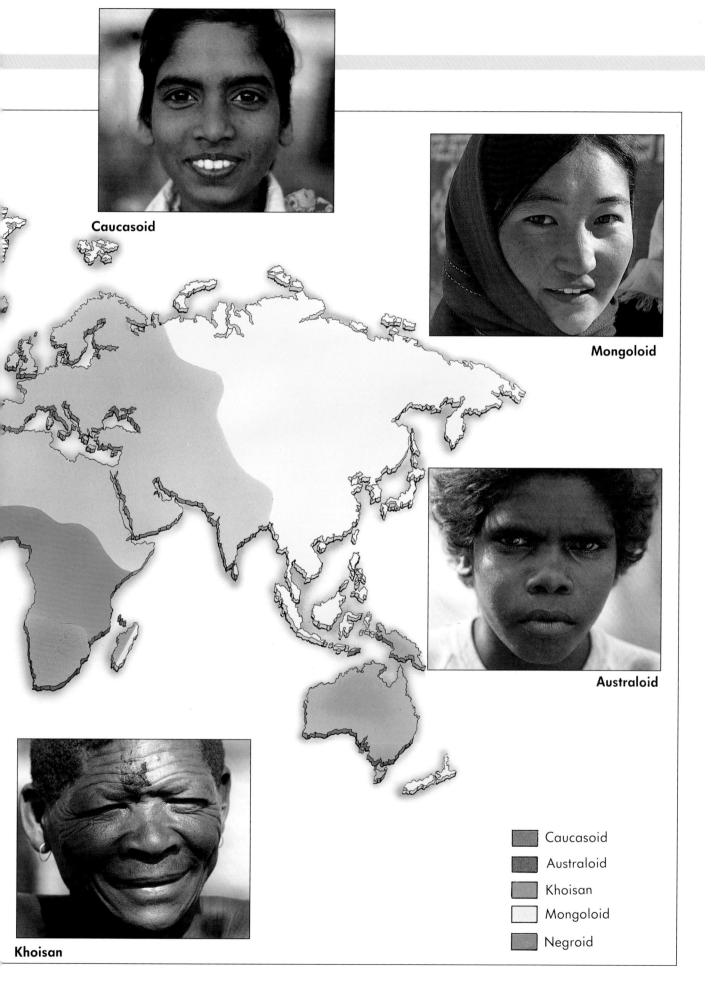

Caucasoid

Mongoloid

Australoid

Khoisan

Caucasoid
Australoid
Khoisan
Mongoloid
Negroid

111

Survival of the fittest

Spot facts

• The fossil record – the remains of early creatures found in rocks – dates from about 600 million years ago.

• The oldest species of living creature is a kind of shellfish that evolved into its present form about 550 million years ago.

• The ancestors of modern man did not appear on Earth until about 4 million years ago.

• Sabre-toothed tigers did not become extinct until 11,000 years ago.

• It takes at least 300,000 generations for a new species to evolve from an existing one.

• It is estimated that there are more than 30 million species of plants and animals on Earth today.

▶ Giraffes feeding on leaves in the treetops. According to Darwin's theory of evolution, giraffes have survived as a species because of their long necks. When food is hard to find at ground level, giraffes can continue to feed on high branches. Shorter species cannot reach this food and so they starve.

In the middle of the last century an English naturalist named Charles Darwin put forward a theory that revolutionized scientific thinking. It was a new theory of evolution, in which he tried to explain why and how life on Earth developed, or evolved in the way it did. His ideas still form the cornerstone of the modern theory of evolution. All living things, Darwin said, are locked in a continual fight for survival. Those species and individuals that are in some way better equipped for the fight tend to flourish and produce more offspring. Those that are poorly equipped tend to die out.

Darwin

Early last century, it was widely accepted that all the plant and animal species in existence had been created at the same time. The date of creation had even been calculated as 4004 BC, based on the Bible. But fossils were providing evidence of animals that no longer existed. The first dinosaur fossils were discovered in England in 1822.

In 1859 Charles Darwin challenged the accepted view of creation in one of the most important science books ever written. It had a long and rambling title: *On the Origin of Species by Means of Natural Selection, or the Preservation of Favoured Races in the Struggle for Life.* In the book, Darwin put forward the theory that life on Earth was the result of millions of years of development, or evolution.

As an idea, evolution was not new. In 1806 the Frenchman Jean-Baptiste de Lamarck had proposed that animals evolved because they tried to improve themselves. A giraffe, said Lamarck, grew its long neck by continually stretching up to reach high branches. His theory was soon replaced by Darwin's theory.

Darwin was not a trained scientist – in fact he was going to be a clergyman. But he was very interested in many different sciences, from geology to the breeding of farm animals. In 1831 he embarked on an epic five-year voyage around the world in HMS *Beagle*. It was during this voyage that he gained an insight into the processes that give rise to evolution – adaptation and natural selection. They formed the basis of his theory.

▼ Charles Darwin (left) and Alfred Russel Wallace (right). Both worked out the theory of evolution at the same time. But Darwin published first and got the credit because he had started his work earlier.

▼ Darwin and one of his "relatives", a cartoon from the *London Sketch Book* of 1874. It refers to Darwin's apparent suggestion that we were descended from monkeys. This suggestion, and indeed the whole theory of evolution, brought Darwin into conflict with many people, including the established Church. In fact, Darwin had been very careful to say nothing about human evolution in his book, because he did not want to cause offence. But this did not stop the Church's supporters from ridiculing his ideas.

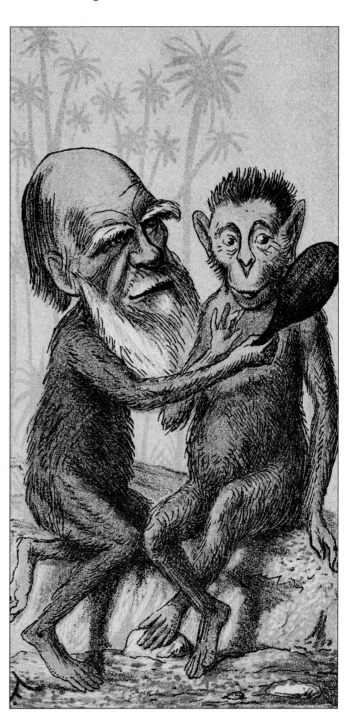

Adaptation

Variety, as Darwin observed, is the essence of life. Butterflies of the same species often have slightly different markings. At the same time, farmers are able to breed sheep and cattle to bring out a desirable characteristic, such as longer wool or more milk. These characteristics are then passed on to successive generations. How does this work in nature?

Life in the wild is at the mercy of the environment. In a species of millions of individuals, each new generation has some individuals with slight differences in physical form or behaviour. Many of these differences are useless, but some may prove to be an advantage. Changes that are beneficial to the species are called adaptations. Through many generations, an adaptation tends to spread throughout the species in a particular area. As a result, the species becomes well-adapted to the local environment. But the species has changed. The new characteristics mean that it has become different. In other words, it has evolved.

▼ A Giant panda feeding on its favourite food, bamboo. Its natural habitat is the foothills of the Himalayan Mountains, where bamboo is one of the commonest plants. The panda can live exclusively on bamboo. It has adapted physically for this diet by developing a sixth digit on its forepaws. This works like a thumb and enables the panda to grip the bamboo.

The more complex an organism is, the more scope there is for variation and adaptation. For example, the early shrimp-like arthropods that lived in water already had legs and eyes. They were therefore well-adapted to make the move from sea to land. Some have remained very much as they were. They still exist as scorpions. Others adapted to their new environment by evolving, over millions of years, into insects or spiders. Once crawling insects were established, some adapted further. They developed wings and took to the air.

The slow process of adaptation enables each species to find its own particular place in the great web of life. Adaptation sometimes causes what is known as convergence of form. This means that species that have similar breeding, feeding and other habits, such as sharks and dolphins, tend to look like each other. On the other hand, adaptation also causes some species to become very specialized, so that they can exist only in very special conditions.

▼ A desert grasshopper rests on the stony ground near two plants called lithops, or living stones. Both species are well camouflaged. Camouflage is a form of adaptation found widely in nature. By taking on the colours of the natural background, the grasshopper becomes much more difficult for a predator to see. And the lithops are less likely to be eaten.

▲ A dense flock of wrybills takes to the air from a seashore in New Zealand. One advantage of living in a flock is that there are many eyes available to spot predators. And if an attack does take place, the attacker will be confused by the mass of fleeing birds. So a bird in a flock has a better chance of survival than if it lived alone.

◄ This insect is a kind of mantis. It is an example of the most extreme form of adaptation. As soon as an insect lands on the "flower", it is in the insect's grasp. But the mantis is over-specialized. Without the orchid its adaptation is useless, because its strange shape and bright coloration stand out clearly against any other type of vegetation.

Natural selection

The process of evolution has been going on continuously for millions of years. We might therefore expect the world to be populated by thousands of millions of different species. But it is not. The latest estimate is about 30 million species. In other words, as evolution proceeds, many new species appear, but also many old species die out, or become extinct. What causes this to happen? What process determines whether a species survives or not? Darwin called the process natural selection.

The most important factor in the struggle for survival is the availability of food. Competition for food is usually strongest between members of related species that have similar feeding habits. Whenever food is in short supply, certain individuals may be better adapted to survive. For example, animals with a long neck can reach leaves on high branches when all the leaves on lower branches have been eaten. These animals are therefore able to survive after animals with short necks have died out. In this way, natural selection preserves any useful adaptations, such as a long neck, that may occur. Over thousands of generations these differences become established in the species and are passed on from parents to offspring.

Adaptation tends to encourage lifeforms in the same habitat to become similar in shape and behaviour. Natural selection, on the other hand, encourages the divergence of characteristics. If all organisms were the same size and ate the same food, then a single change in the environment could wipe out all life. Natural selection makes sure that there are always some forms of life able to survive.

Adaptation and natural selection form the basis of the theory of evolution. Working over millions of years, adaptation causes constant change and variety among species. At the same time, natural selection ensures that only the fittest species survive.

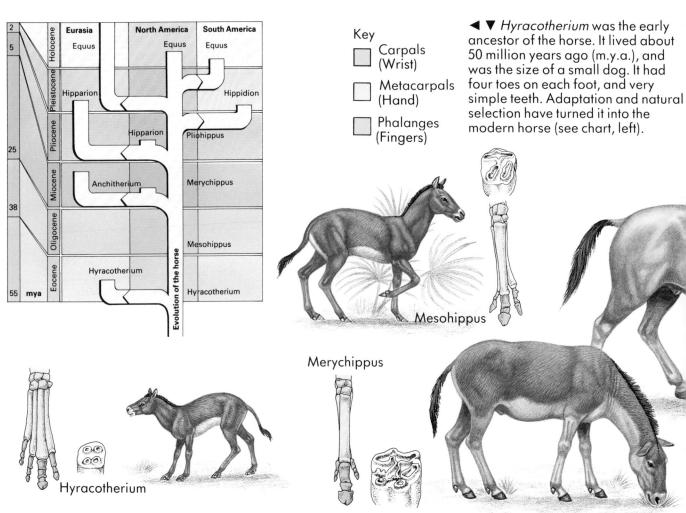

Key
- Carpals (Wrist)
- Metacarpals (Hand)
- Phalanges (Fingers)

◀ ▼ *Hyracotherium* was the early ancestor of the horse. It lived about 50 million years ago (m.y.a.), and was the size of a small dog. It had four toes on each foot, and very simple teeth. Adaptation and natural selection have turned it into the modern horse (see chart, left).

Mesohippus

Merychippus

Hyracotherium

Evolution of the horse

mya		Eurasia	North America	South America
2 / 5	Holocene	Equus	Equus	Equus
	Pleistocene	Hipparion		Hippidion
25	Pliocene	Hipparion		Pliohippus
	Miocene	Anchitherium		Merychippus
38	Oligocene			Mesohippus
55	Eocene	Hyracotherium		Hyracotherium

Living fossils

In 1938 men fishing in the Indian Ocean off the coast of Africa netted a fish that had never been seen before. It was later identified as a kind of lobe-finned fish which was thought to have become extinct long ago. It was a coelacanth. A second coelacanth was caught in 1952, and since then, more than 80 have been caught. Recently, other coelacanths have been filmed in their natural deep-water habitat.

The coelacanth is a close relative of the type of fish that evolved into amphibians during the Devonian Period, about 350 million years ago. The discovery of the coelacanth was a reminder of just how incomplete the fossil record is.

▶ This is a fossil of *Hyracotherium*, the ancestor of the horse. Over a period of 50 million years, the horse has grown progressively larger. Its feet have gradually evolved from having four toes to only one, because a single hoof on each foot is better for galloping at speed over open grassland. Also, the ridges on the surface of the teeth have become much more complex. The ridges make the teeth much more efficient at grinding up grass.

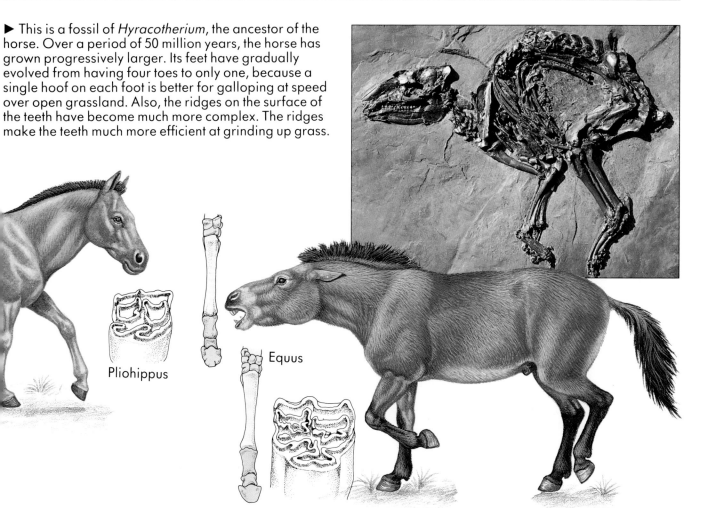

Pliohippus

Equus

Evolution in action

The theory of evolution explains why living things developed the way they did. It is almost certain that plants developed first, and that the first multi-celled animals were herbivores, or plant-eaters. Meat-eating animals, or carnivores, developed much later, when there were enough herbivores for them to feed on. Depending on the animals concerned, it takes 20 to 100 herbivores to support one carnivore. For example, a lion that feeds once a week must kill and eat 52 animals every year.

Adaptation is constantly producing new types of herbivore, which are more efficient at feeding on the available vegetation. At the same time, new forms of carnivore are developing, which have to compete with each other to feed on the available prey animals. In both cases, natural selection produces specialists and generalists. Specialists are animals that feed on one particular type of plant or animal. Generalists are animals that are much less choosy about their food.

Varieties of marsupials. Marsupials are mammals that raise their young in a pouch. They evolved separately from other mammals. Yet adaptation to similar habitats has produced similar body forms in marsupials and other mammals. The wolf (1a) and marsupial wolf (1b) are medium-sized predators. The wood mouse (2a) and marsupial mulgara (2b) are small scavengers. The Californian mole (3a) and marsupial mole (3b) both live underground.

◀ Dark and light forms of the Peppered moth on a soot-covered tree. The Peppered moth originally had a pale, speckled appearance. This camouflaged it from predators while it was resting on a tree branch. During the Industrial Revolution, smoke and soot turned many trees almost black with pollution. As a result, the Peppered moth developed a much darker coloration, which provided much better camouflage in the new conditions.

This new form of the moth remained confined to industrial areas. In country areas, far away from industrial cities, the moth still kept its paler markings. Now that the industrial nations are starting to control air pollution, the darker varieties of moth will probably adapt back to their original colour.

Predators and prey

The relationship between animals that hunt (predators) and animals that are hunted (prey) has caused some of the most interesting of all adaptations. Usually, prey animals rely on speed to escape from predators. But sometimes evolution takes a different course. The struggle for survival becomes a straight battle between attack and defence.

This evolutionary "arms race" was notable when dinosaurs ruled the Earth. Despite its fearsome appearance, the 5-tonne *Triceratops* (below left) was a placid herbivore. The armoured neck frill and razor sharp horns were purely for defence. *Triceratops* evolved such massive protection because it shared its habitat with some very fierce predators. *Tyrannosaurus* (below right), for example, was superbly equipped as a hunter. It had a massive head and powerful jaws.

But it probably left *Triceratops* well alone, because the chances of a successful attack were greatly outweighed by the risk of serious injury.

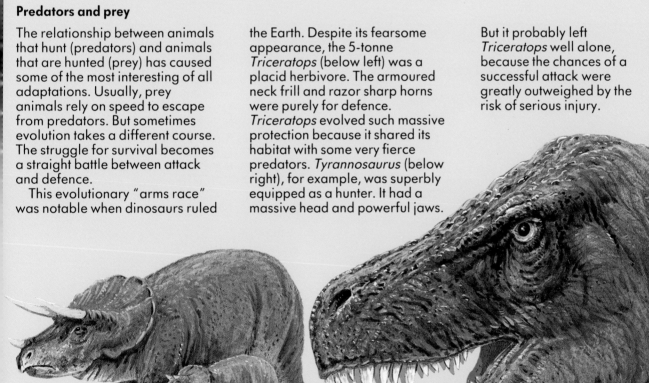

Evolution and the changing Earth

During the time life has flourished on Earth, the shape of the Earth's surface has changed completely. The change has been brought about by continental drift. This is the very gradual movement of plates, or sections, of the Earth's crust. In the Cambrian Period there were only four continents. They were not the continents we know today, and they were in very different positions. Gradually they drifted together. By the time of the Devonian Period there were only three continents.

By the Permian Period, all the continents had come together to form one super-continent, known as Pangaea. By the middle of the Cretaceous Period, the present-day continents were drifting apart. During the Tertiary Period they moved towards their present positions. By about 40,000 years ago, the world was almost exactly as we know it today. Yet the continents are still moving very slowly. In 100 million years time the world will again look different.

The drifting of the continents and other changes in the land masses made the sea level rise and fall. While life remained in the seas, changes in sea level made little difference to evolution. But once life had moved on to the land, the changes in sea level started to affect the course of evolution. For example, during the Devonian Period, the sea level was very low and drought was widespread. But during the following Carboniferous Period, the sea level was much higher. This caused the swampy and humid conditions in which giant forests and amphibians thrived.

Later, the various land masses became separated. This allowed evolution to continue in isolation. During the last 5 million years, ice ages have also affected the distribution of animals. For example, they caused bridges of frozen sea to form between Asia and North America. Animals migrated between the two continents across the bridges.

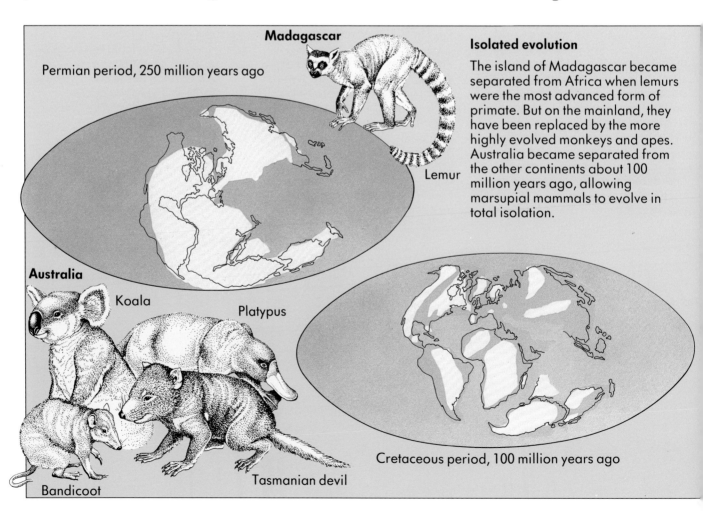

Madagascar

Permian period, 250 million years ago

Lemur

Australia

Koala

Platypus

Bandicoot

Tasmanian devil

Isolated evolution

The island of Madagascar became separated from Africa when lemurs were the most advanced form of primate. But on the mainland, they have been replaced by the more highly evolved monkeys and apes. Australia became separated from the other continents about 100 million years ago, allowing marsupial mammals to evolve in total isolation.

Cretaceous period, 100 million years ago

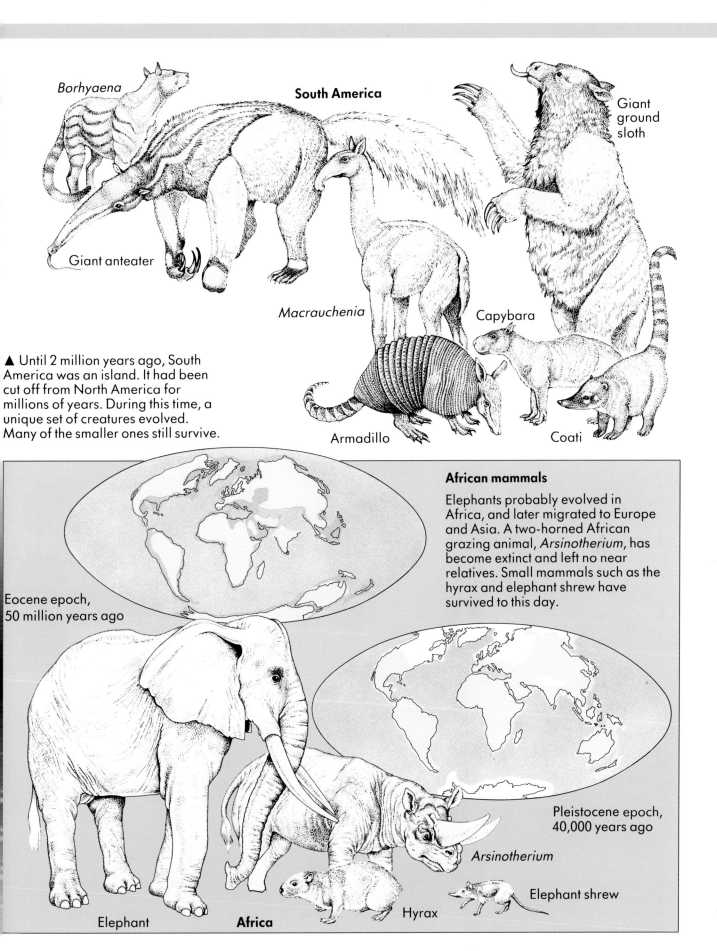

Borhyaena

South America

Giant ground sloth

Giant anteater

Macrauchenia

Capybara

▲ Until 2 million years ago, South America was an island. It had been cut off from North America for millions of years. During this time, a unique set of creatures evolved. Many of the smaller ones still survive.

Armadillo

Coati

African mammals

Elephants probably evolved in Africa, and later migrated to Europe and Asia. A two-horned African grazing animal, *Arsinotherium*, has become extinct and left no near relatives. Small mammals such as the hyrax and elephant shrew have survived to this day.

Eocene epoch, 50 million years ago

Pleistocene epoch, 40,000 years ago

Arsinotherium

Elephant shrew

Elephant

Africa

Hyrax

121

Lowly life

● *One gram of soil can contain up to 2,500 million bacteria.*

● *In good conditions some bacteria divide in two every half hour. In one day a single bacterium could become 281 million million.*

● *The giant kelp seaweed can grow 60 m long – and it only lives for a year.*

● *Some fungi make new threads 1 km long in one day.*

● *"Red tides" occur when a sea is coloured by vast clouds of tiny red algae. Some contain powerful nerve poisons. Some are so strong that one gram would kill 5 million mice.*

▶ The soil fungus *Rhizopus.* From tangled threads rise stalks tipped with rounded spore cases. When ripe, these burst and scatter thousands of spores. This fungus is similar to the pin-mould *Mucor*, which may be found in kitchens living on damp bread or decaying vegetables: Many fungi live on dead plant and animal remains.

For at least 3,000 million years there have been living things on the Earth. The earliest two kinds known, bacteria and blue-green algae, are still common and important today. But they remain tiny, and we can only see what they are like using microscopes. Each consists of just a single cell. The simplest animals are the protozoa. These too consist of just a single cell. Most living things are made up of many cells. Amongst the algae we can see both single-celled and many-celled forms. More complex animals and plants have developed as time has gone on.

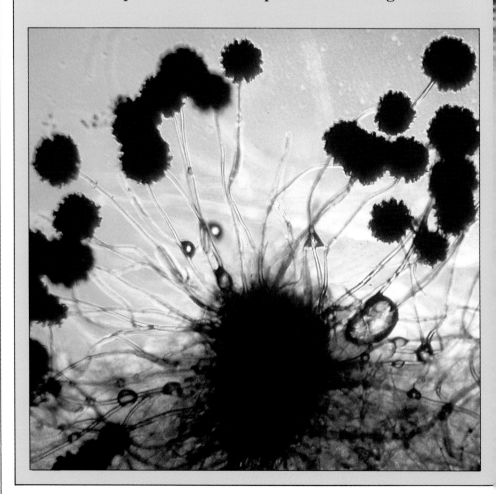

The microbes

Bacteria and blue-green algae have a cell wall but no nucleus or specialized parts within the cell. Bacteria live almost everywhere – in soil, air, ocean depths, even in hot springs at 92°C. Some use sunlight to make food as plants do. Others make energy from other chemical reactions, such as turning sulphur to sulphuric acid. Many are useful to us and other living things. Bacteria trap atmospheric nitrogen and turn it into substances that act as fertilizers for plants. Bacteria also cause many diseases, such as cholera and tetanus.

Single-celled animals have a nucleus and a soft cell membrane. Protozoa like the familiar amoeba live in fresh water; some live in the sea and soil. Some are parasites that cause diseases such as malaria and sleeping sickness.

Viruses cannot survive by themselves, but are found inside the cells of other living things. There they make copies of themselves. They cause a variety of diseases.

▼ The sharp point of a pin (1) is magnified 175 times by a scanning electron microscope. The tiny rod-like shapes clustered in the lower part are bacteria. At much greater magnification (2) is an electron microscope view of a virus. This is *Coronavirus*, one of several viruses that produces human symptoms known as the common cold. The grape-like bunches (3) seen by electron microscope are rounded, or coccus bacteria. These are *Staphylococcus aureus*, which usually live harmlessly on human skin. If these bacteria get into the body through scratches and cuts they may cause boils and other infections.

3

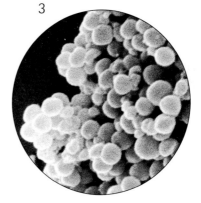

2

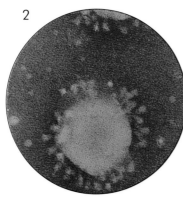

1

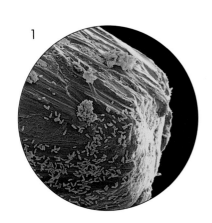

▼ Some fungi can be seen by the naked eye. Smaller fungi, and the protozoa, need a light microscope to be seen well. Bacteria may be visible in a light microscope, but details need the greater magnification provided by an electron microscope. Viruses are so small that they can only be seen with the help of an electron microscope.

Magnifications

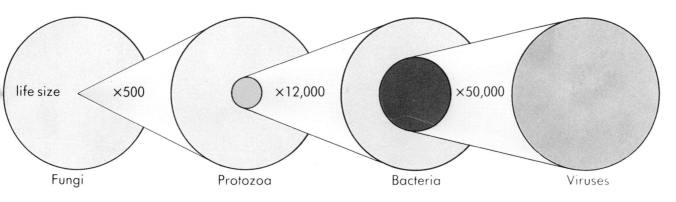

life size   ×500   ×12,000   ×50,000

Fungi   Protozoa   Bacteria   Viruses

Algae

Algae are simple plants that do not have true stems, leaves or roots. Many of them are single-celled. These include the diatoms, which have a shell like a carved pill-box. They live drifting in the sea or fresh water, and are food for many animals. Dinoflagellates also have shells round their single cells. They live in the sea, and have two flagella. These threads stick out from the cell wall and beat, spinning them as they move through the water. *Euglena* and its relatives live in fresh water. *Euglena* has a single flagellum.

Euglena is green, and uses sunlight to make its own food. If there is no light it can absorb foods from the water. Some of its close relatives have no green pigment and, like animals, feed on other living things. The larger algae are green, brown or red. Many seaweeds seen on the shore are brown algae. Most seaweeds are red algae. They can live in relatively deep water. Algae reproduce in various ways. Some just divide; others make tiny sex cells that join to form a new generation. Some reproduce by shedding spores into the water.

The variety of algae

Algae show an enormous range of sizes and shapes, almost as though they had experimented with every way of putting cells together to make a larger body. Shown here are just a few of the green algae. *Spirogyra* (photo below) is an alga that forms a green scum on fresh water. The microscope shows that its filaments are made of cells joined end to end. It can be recognized by its chloroplast which forms a spiral. *Staurastrum* (1) lives floating in lakes. It is single-celled and the cell wall is spiky. In *Eudorina* (2), a freshwater form, 32 cells live together in a ball of jelly. Species of *Cladophora* (3) are found in fresh water and on the seashore. Attached at one end, the filaments of cells branch and have a wiry feel. Some grow 1 m long. *Acetabularia* (4) lives on rocks on the lower shore and consists of a single cell as long as your finger. *Codium* (5) is a seaweed about 30 cm long. It is made up of a mass of filaments interwoven into branching tubes with a felt-like feel. The sea lettuce, *Ulva* (6), has fronds with a double layer of cells.

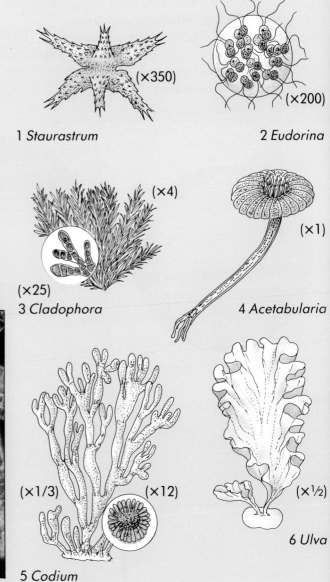

(×350)

1 *Staurastrum*

(×200)

2 *Eudorina*

(×4)

(×25)

3 *Cladophora*

(×1)

4 *Acetabularia*

(×1/3) (×12)

5 *Codium*

(×½)

6 *Ulva*

Fungi

Fungi differ from true plants in having no green pigment and not making their own food. They need to get food from outside their bodies. Many are parasites of plants, and include serious pests of crops. Some live on the bodies of live animals. Large numbers get their food by breaking down the remains of dead animals or plants. Some form associations with plants, especially algae, from which both benefit.

There are about 100,000 species of fungi.

Some, like the yeasts that ferment and help us make beer, wine and bread, are single-celled. But many fungi are much larger. Often the main part of a fungus remains unseen, and consists of a mass of threads underground. We notice the fruiting bodies that grow from these threads because they grow above ground. They release spores in thousands which are carried through the air. Some find a suitable spot to grow. Some fruiting bodies are small, and rarely seen. Others are the familiar mushroom and toadstools.

◀▼ It is on the fruiting bodies of fungi that spores are produced. A variety are shown below. In the puffball (photo) the spores are puffed out through the hole on top when it is disturbed.

Mushroom life-cycle

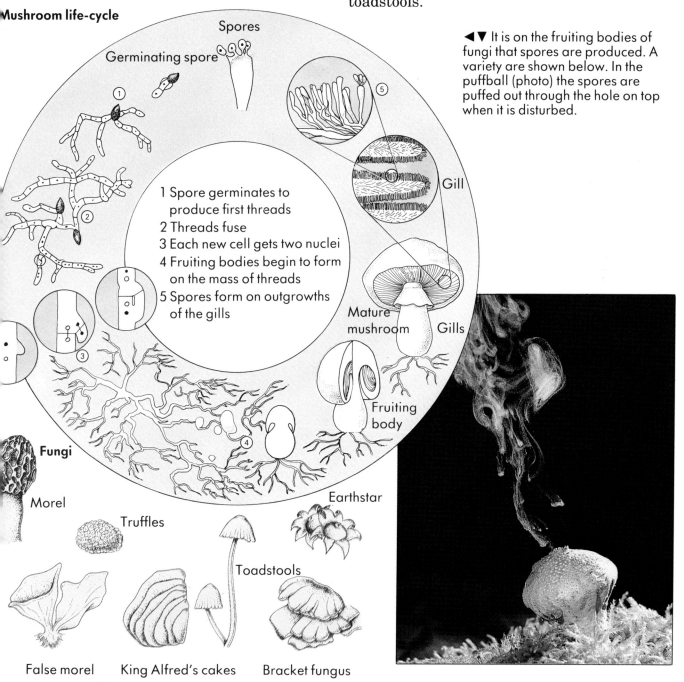

Spores

Germinating spore

1 Spore germinates to produce first threads
2 Threads fuse
3 Each new cell gets two nuclei
4 Fruiting bodies begin to form on the mass of threads
5 Spores form on outgrowths of the gills

Gill

Mature mushroom

Gills

Fruiting body

Fungi

Morel

Truffles

Earthstar

Toadstools

False morel   King Alfred's cakes   Bracket fungus

125

Liverworts, mosses and ferns

Mosses and liverworts are low-growing plants that often form mats or cushions. They are found mainly in damp and shady places, although some mosses can stand being dry for months. They have no true stems or leaves, nor do they have proper roots. They absorb water and foods from below through hair-like cells. They have no special tubes for carrying water and food, and this limits their size.

Mosses and liverworts have two stages to their life-cycles. One stage produces male and female cells. These unite to produce the next stage, which is a spore-bearing plant. The spores give rise to a new sexual generation. Liverworts get their name from their sexual stage. This is often a flat but fleshy plant body shaped like a liver. Other liverworts have flattened "leaves" growing from a central body. Mosses always have leaves carried on a midrib. The spore-bearing stage of these plants grows on the sexual stage. It consists of a foot, a stalk, and a capsule in which spores develop.

Ferns also have sexual and spore-bearing stages. The sexual stage is small, short-lived and rather like a flat liverwort. The familiar large ferns such as bracken are spore-bearing stages. Under the fronds you can find brown patches, which are where the spores form. Ferns have tubes to take water and food through their bodies, and can grow much larger than mosses. They have roots and leaves which are often complex fronds. These are able to catch enough light even in shady places. Like mosses and liverworts, ferns are commonest in warm countries. Ferns are most abundant in South-east Asia. A few grow tree-sized.

Horsetails have an upright stem with rings of side branches. Few species of these ancient plants survive. The spore-bearing parts form "cones" at the tip of the shoots. Like ferns, they eject their spores clear of the plant.

Although most are not conspicuous there are about 9,500 different kinds of moss, 6,000 liverworts and 12,000 ferns.

Moss life-cycle

▶ Spore capsules grow from a moss plant (1) and (2). When mature, their caps come off, exposing a ring of "teeth" (photo below). This opens in dry air and lets out spores (3). A spore grows into a green thread (4). Moss plants bud from this (5). Male parts (antheridia) and female parts (archegonia) develop on the plant (6). Male cells swim to fertilize the egg (7) which divides (8) to form a spore capsule.

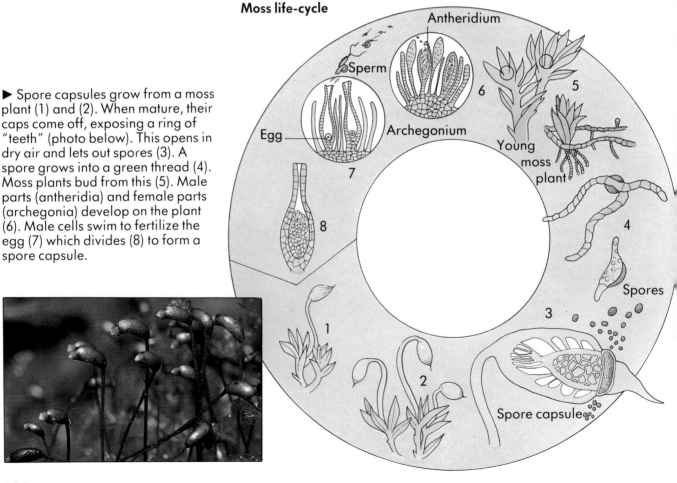

126

▲ A damp southern beech forest in New Zealand. Here mosses, ferns and lichens flourish, many of them growing on the trees. Ferns and mosses need a film of water for their male cells to swim through in order to reach the female cells. Damp conditions, as here, are ideal for reproduction. Ferns and mosses thrive in such wet places, but cannot succeed in drier areas, such as deserts.

◄ The soldier lichen is two plants combined: an alga and a fungus. Often the same sort of alga also lives free, but the fungus needs the alga to live. One way lichens reproduce is by throwing off balls of fungus threads enclosing algal cells. These bundles give a dusty look to the soldier lichen. An association such as this, where two species combine, is called symbiosis.

▼ A lichen is made of an alga cell surrounded by fungus threads.

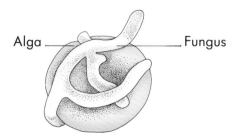

Alga _____ Fungus

Plants and pollen

▶ A bee visits a flower and picks up some pollen. Once plants developed pollen, with its waterproof covering, they no longer needed damp conditions to reproduce. The male cells in pollen could be carried on the wind or by insects. Flowers pollinated by insects often have colours or markings which attract the insects.

Pollen is a type of plant spore. It germinates to produce male cells. It gives a good means for dispersal, even in dry conditions. The first plants to produce pollen were the gymnosperm, or "naked-seed" plants. Another type of spore, which develops into female cells, grows on the plant still inside its "spore case". Once male cells from pollen fertilize the female cell, it develops into an embryo still inside this case. The embryo with its case is called a seed. In gymnosperms the seed has no other cover. In flowering plants, which are called angiosperms, the seeds are wrapped in a further layer. The whole thing is known as a fruit. Plants have many ways of dispersing their seeds.

Primitive gymnosperms

The first of the gymnosperm group of plants lived over 220 million years ago, before the time of the dinosaurs. They had developed two big advantages over plants that had gone before. They had pollen that could be carried long distances, and they had seeds. In a seed an embryo plant can remain dormant until conditions are right for it to grow.

The first seed plants we know of were rather fern-like. Later they grew as large trees. Most of the living gymnosperms are conifers such as fir trees. But there are still a few examples of old-fashioned types of gymnosperm in isolated parts of the world.

The cycads have spreading palm-like leaves forming a crown on a unbranched trunk. The trunk has a large soft pith. Some cycads are quite short. Others grow up to 20 m high. Long ago they were common, but now only about 100 species survive. Spores are formed in huge cones on the top of the trunk.

There is only one kind of ginkgo, although we know about others through fossils. It grows as a tall narrow tree up to 28 m high. It is planted in cities because it is resistant to pollution.

The gnetophytes are another small group of odd seed plants. They include the shrubby *Ephedra*, from which the drug ephedrine can be extracted. Oddest of all is *Welwitschia*, a rare plant from Namibia in southern Africa. It is a desert plant, good at preventing the loss of precious water. It is woody and has just two strap-like leaves. These grow throughout the life of the plant, and may curl and split to produce a large twisted plant.

▼ This cycad grows in Mexico. It has edible seeds. Cycads look rather like palms, but belong to a separate group of plants more closely related to conifers. They were most successful 150 million years ago, but a small number of species still live scattered through the tropics. The stems of some can be used to make sago.

The living fossil

Leaves and catkins of a ginkgo tree. The ginkgo seems almost unchanged from those that grew 200 million years ago. Now it grows wild only in a remote part of China. Pollen is produced on the catkins and carried by the wind. The female parts sit at the ends of special shoots.

The conifers

There are about 550 species of conifer. They are most common in the cooler parts of the world. Some live at the limits of tree-growing zones on mountains or towards the poles. Conifers are mostly upright plants and are woody shrubs and trees. They include the bulkiest and tallest plants on Earth, the giant redwoods.

The cones that give these trees their name come in two forms, male and female. These are usually on the same tree but on different shoots. The male cones are usually small, but the mature female cones can be woody structures up to 40 cm long in some species. A few species, such as junipers and the yew, produce a fleshy cone. The red yew "berry" is attractive to birds. This helps distribute seeds, as the birds carry the seeds away from the tree.

Conifers usually have needle-like leaves. Some, such as cypresses, have short scaly leaves round their shoots. The monkey-puzzle tree from South America has huge spiky scales. Needles have a hard outer skin and are a good shape for retaining water. Conifers can live in dry conditions, as in the cold northern winter, or on poor sandy soils. Most conifers keep leaves all the year, but larches shed them in winter.

Conifers can form dense forests. Because they let little light through to the ground, these forests may have little undergrowth. But many fungi live in the ground in association with their roots. Some appear as toadstools.

Conifers grow fast in cool climates and on soils that are no good for other crops, so they are much used in forestry. Their wood, called

softwood, can be used as timber and for papermaking.

The Norway spruce is familiar as the Christmas tree. Europe's tallest native tree at up to 54 m, it may be 30 years old before it starts making cones. The age of many conifers can be judged by the number of rings of branches. There is usually one for each year. But in species such as the Scots pine the older trees often lose their lower branches.

◀ The magnolia is an ancient type of flower. Scientists believe that all flowers developed from this type. It has female parts, the carpels, and male parts, the stamens, as well as petals and sepals in large numbers, which vary from flower to flower. In more advanced plants there may be fewer parts. Some may be lost, or fused to make complex new structures, as in orchids.

Parts of a flower

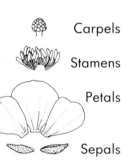

Carpels

Stamens

Petals

Sepals

A pine shoot

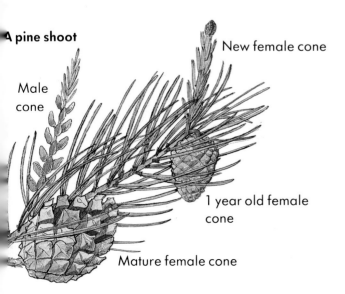

New female cone

Male cone

1 year old female cone

Mature female cone

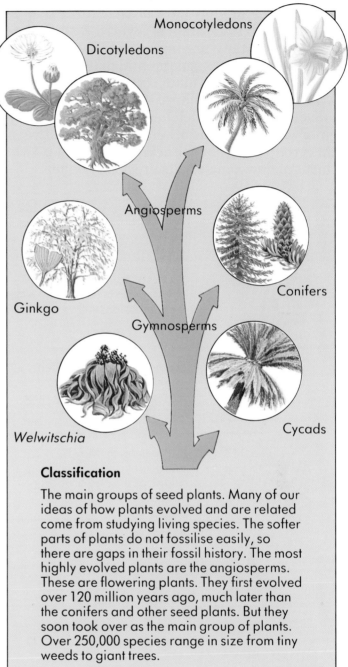

Monocotyledons

Dicotyledons

Angiosperms

Ginkgo

Gymnosperms

Conifers

Welwitschia

Cycads

Classification

The main groups of seed plants. Many of our ideas of how plants evolved and are related come from studying living species. The softer parts of plants do not fossilise easily, so there are gaps in their fossil history. The most highly evolved plants are the angiosperms. These are flowering plants. They first evolved over 120 million years ago, much later than the conifers and other seed plants. But they soon took over as the main group of plants. Over 250,000 species range in size from tiny weeds to giant trees.

◀ (far left) The Monterey pine is found in just one small coastal area of California. This isolated population is descended from pines left behind as conifers retreated north after the last Ice Age.

◀ In conifers, reproduction is a slow process. In the pine, winged pollen from male cones is carried on the wind to a young open female cone. It may take a year for the pollen to germinate and grow to fertilize the egg cell. The female cone closes up after pollination. Seeds mature slowly, and a female pine cone may be three years old before it finally releases the ripe seeds.

131

The flowering plants

Nine-tenths of that part of the world that is covered with vegetation is covered with flowering plants. Many are important to us for food, drugs, timber or as ornamental plants.

There are two main types: dicotyledons, which start life as embryos with two leaves, and monocotyledons, which have a single leaf in the embryo. Some of the other differences between the groups are shown below. More than two-thirds of flowering plants are dicotyledons.

The dicotyledons include many small flowers such as primroses and violets, shrubs such as hazel, and trees such as elm and oak. The monocotyledons are mostly low-growing. They include the grasses, lilies, orchids and palms.

Some flowering plants are pollinated by wind. Most have their pollen taken from flower to flower by insects. The flower itself, with its colours and its nectar, attracts the insects which do this important work for the plant.

Dicotyledons		Monocotyledons	
	Embryo: Two seed leaves. Endosperm (extra food) present or absent.	Embryo: One seed leaf. Endosperm often present.	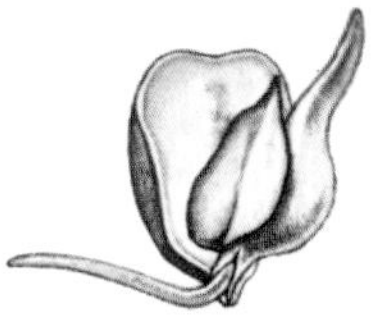
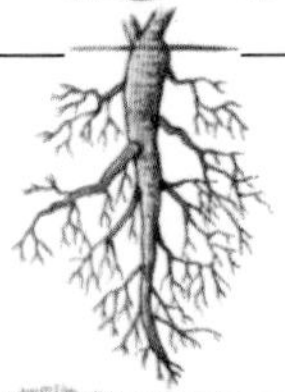	Roots: The first root often persists and becomes a strong taproot, with smaller secondary roots.	Roots: The first root soon disappears, to be replaced by a branched fibrous root system.	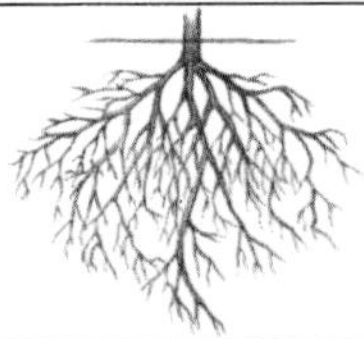
	Growth form: May be a herb, that is, a low-growing plant, or may be a woody shrub or tree.	Growth form: Most are herbs. A few, such as palms, are tree-like.	
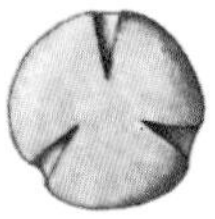	Pollen: Usually has three furrows or pores.	Pollen: Usually a single furrow or pore.	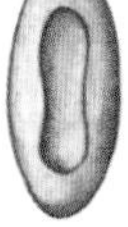
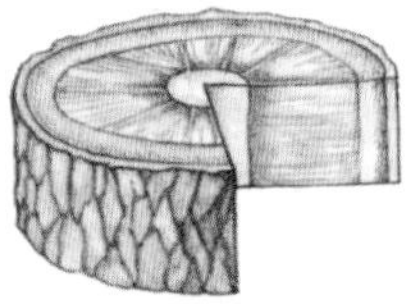	Vascular system: Tubes for conducting water and food are in a ring round the outside of the stem.	Vascular system: Tubes for conducting water and food are scattered across the width of the stem.	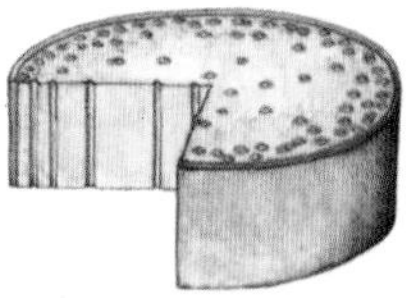
	Leaves: Usually broad, and with the veins forming a network. May be divided or compound.	Leaves: Usually long and narrow, often sheathing the stem. The veins run parallel to the long axis.	
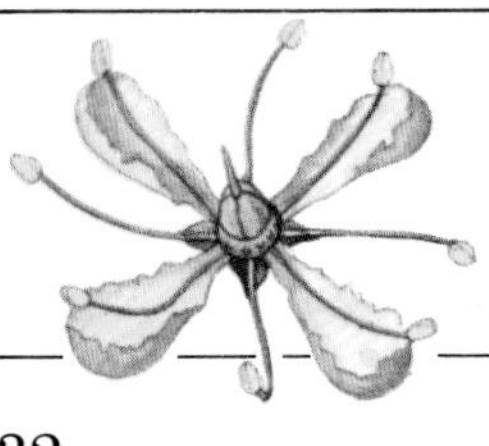	Flowers: Parts are typically arranged in fours or fives.	Flowers: Parts are usally arranged in threes or multiples of three.	

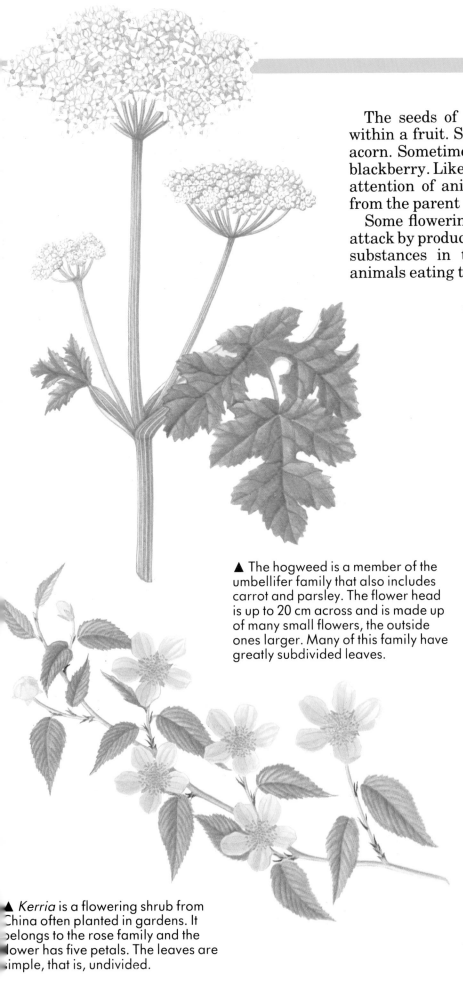

The seeds of flowering plants are enclosed within a fruit. Sometimes this is hard, like the acorn. Sometimes it is fleshy, like the plum or blackberry. Like flowers, fruits often attract the attention of animals which carry seeds away from the parent plant to grow.

Some flowering plants are protected against attack by producing strong-tasting or poisonous substances in their leaves. These can stop animals eating them.

▲ The hogweed is a member of the umbellifer family that also includes carrot and parsley. The flower head is up to 20 cm across and is made up of many small flowers, the outside ones larger. Many of this family have greatly subdivided leaves.

Meat-eating plants

A Lesser bladderwort catches a mosquito larva. The small bladders on the leaves of this water plant are traps for tiny prey. They have inward opening trap doors, set off by trigger hairs on the outside. The prey is pulled in with the rush of water. It dies and decays. It then provides the plant with nourishment. This is usually lacking in the water of the peaty pools where the plant lives. Other insect enemies among plants include sundew, which catches them on sticky leaves, and the Venus' fly trap, in which the leaves snap shut to catch insects. All these plants live on poor soils.

▲ *Kerria* is a flowering shrub from China often planted in gardens. It belongs to the rose family and the flower has five petals. The leaves are simple, that is, undivided.

Animals without backbones

• *The largest animal without a backbone is the Giant squid. It can reach 20 m in length.*

• *Millions of termites may live in a single nest. Some nests are 7.5 m tall.*

• *Over 200 million people are infected with the Blood fluke Schistosoma.*

• *The 1,900-km long Great Barrier Reef off the coast of Australia is the largest structure built by living things. It is made of coral.*

• *A dragonfly's eye has 30,000 facets.*

The earliest fossil animals known, from 650 million years ago, are soft-bodied jellyfish. By 600 million years ago we know there were many other animals, including some with shells or hard skins. But they did not have backbones. We call them invertebrates. Since then, animal bodies have become more complicated and efficient. Better senses and brains have evolved. But many of the 39 main groups, or phyla, that we know today already existed 600 million years ago. Some animals have hardly needed to change since then, perhaps because they fit so well with their surroundings. This group includes the highly successful insects.

▶ Shoals of the colonial jellyfish *Porpita* float in the surface waters of the sea. Like other jellyfish, and their coral and sea anemone relatives, they have stinging cells to capture small animals as food. The sea was home to the first animals. Most types of simple animals still live there.

Classifying the animals

Animals are classified into phyla depending on their body plan. For example, all soft-bodied worms with bodies made of a series of rings, or segments, are put in a phylum called the Annelida. The Mollusca have soft unsegmented bodies protected by one shell or two. Animals in most phyla have a head and a tail. But the Cnidaria and Echinodermata are symmetrical around the vertical axis.

One stage up from the Protozoa, which have a single layer of similar cells, sponges have more than one layer of cells but do not have specialised nerve or muscle cells. The higher animals have basically three layers of cells. These animals have "organs" containing cells which are specialized for a particular job.

Almost all phyla contain species that live in the sea. Fewer species have become adapted to fresh water. Only the jointed-legged animals such as insects and spiders are common on land.

Invertebrates

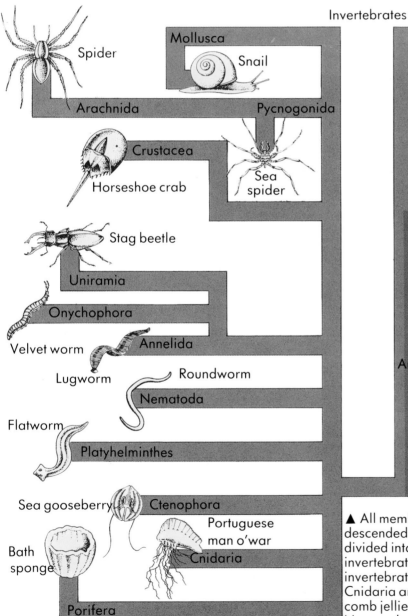

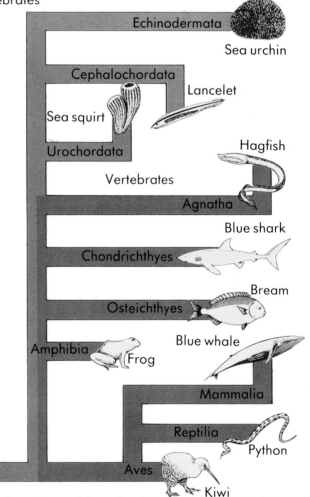

▲ All members of the animal kingdom are probably descended from a single common ancestor. Animals are divided into vertebrates: animals with backbones, and invertebrates: animals without backbones. Amongst the invertebrates (from the left), the Porifera are sponges; Cnidaria are corals and jellyfish. Ctenophora include comb jellies; Platyhelminthes are flatworms and flukes. Nematoda are roundworms and eelworms; Annelida are segmented worms. Uniramia include all insects, centipedes and millipedes; Crustacea are shrimps, crabs and woodlice. Snails, mussels and squid are included in the Mollusca.

135

Simple animals

Sponges live on the seabed. They have a central cavity, and draw water into this through many small holes in their bodies. It is sent out of one large hole at the top. On the way, food is removed. Many sponges have hard crystals of silica in their bodies. The bath sponge has horny fibres instead.

Cnidarians include corals and jellyfish. They have circular bodies. There is a single opening into the body cavity where food is digested. They catch prey using the stinging tentacles around this mouth. Some corals form huge colonies which secrete a stony skeleton to live in. Coral reefs form in tropical seas and take centuries to build up.

Several kinds of animal are called worms. Annelids include earthworms, which do useful work in turning over soil. Many annelids live in the sea. Some burrow, for example lugworms. Others live in tubes. Some walk on the bottom on fleshy "legs". Leeches have suckers and jaws to feed on the flesh or blood of other animals.

Flatworms are another important group. Some glide over the bottom of the sea or fresh water, but many kinds are parasites. These live in other animals, including humans. They include tapeworms and liver flukes. Roundworms have no segments, and are long and narrow with pointed ends. They have very tough skins. Many live in the soil, but some are parasites on plants, animals or humans.

▲ Part of the Caribbean seabed. The leaf-like Sea fan and the Brain coral in the foreground are cnidarians, as is the branching hydroid colony. A flat red encrusting sponge can be seen. Tube-shaped yellow sponges point upwards. None of these animals can move around. They sieve out tiny food particles in the water.

▼ An earthworm is made up of a series of rather similar segments, but the hearts and sex organs are towards the front of the animal. It does not have complex eyes, but simply cells on its back that are sensitive to light. Other cells respond to chemicals or touch. Worms swallow soil as they burrow and digest decaying plant fragments in it. Earthworms burrow in soils all over the world, and need to keep their slimy skins damp.

▼ The fanworm is a segmented worm that lives on the seabed. It builds a tube and lives inside. The head is crowned with a ring of tentacles covered in tiny beating hairs, or cilia. The fan of tentacles is spread from the end of the tube and used as a gill to get oxygen from the water. It also traps tiny pieces of food. These are passed down by the cilia to the mouth.

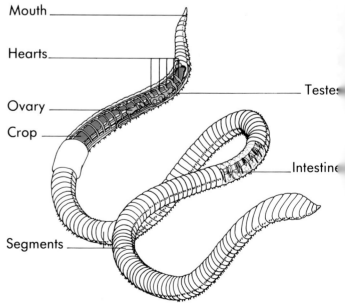

Mouth

Hearts

Ovary

Crop

Testes

Intestine

Segments

The molluscs

There are nearly 80,000 species of mollusc. They include snail-like forms with a single coiled shell; clams and mussels with a pair of shells; and the octopuses and squids, in which the shell is internal or absent. Land snails are usually plant-eaters. Some of the sea snails are meat-eaters. Bivalves such as mussels are filter-feeders and are able to stay in one place and let their food come to them. Squids and octopuses are fast-moving, active hunters. They have efficient bodies, good eyesight and large brains. They are the most intelligent of animals without backbones. Most puff out a blob of "ink" when threatened.

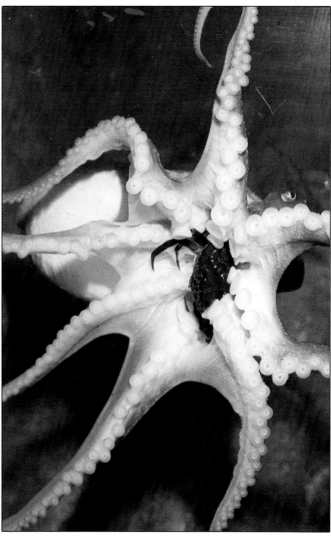

▲ The head of a land snail. The tentacles bear sense organs including eyes. The mouth contains a long tongue covered in sharp teeth. It is used like a file to rasp at the plants it feeds on. Snails are the only molluscs living on land. They glide on a carpet of slime. The shell gives protection from enemies and from drying out.

▲▼ A Lesser octopus catches a crab. The arms grab prey which is killed by a bite from the beak. Octopuses spend most of their time on the sea bottom. Squids and cuttlefish (below) are good swimmers. They take water into the mantle cavity for the gills. They can swim by jet propulsion by shooting water out of the funnel.

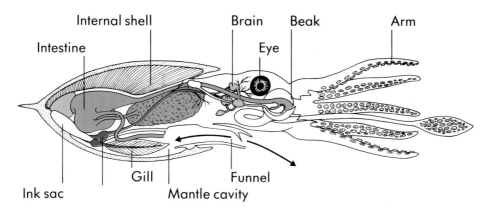

Internal shell   Brain   Beak   Arm

Intestine   Eye

Ink sac   Gill   Mantle cavity   Funnel

Insects

There are more species of insect than all the other kinds of animal put together. Over one million have already been named. Scientists believe there may be even more waiting to be described. They live in almost every place except the sea. Every food seems to have some kind of insect that eats it. What is the secret of their success?

An insect's body is divided into three main parts: the head, thorax and abdomen. The head has eyes and antennae. The thorax has three pairs of legs and usually two pairs of wings. Being able to fly is a big advantage. Insects have a hard outer skin that also acts as a skeleton to support them. This skin is quite waterproof, and helps insects succeed on land.

The disadvantage of such a skin is that it cannot grow. To grow, an insect must shed its skin, and expand while the new one is soft. There are two main ways that insects grow up. Some insects, such as grasshoppers and cockroaches, hatch from eggs as small versions of adults. But they lack wings, and are unable to breed. They moult several times before becoming adult. Other insects, such as butterflies, flies and wasps, hatch from the egg as a wriggling larva quite different from the adult. This feeds, moults and grows. Then it goes into an immobile stage, the pupa, in which its body is totally reorganized. From it, the insect emerges as a fully-formed winged adult.

Some insects such as bees, ants and termites are social animals with highly organized nests.

Some insects damage plants and stored crops. Others carry disease. But many are important to humankind in pollinating crop plants.

The variety of insects. The fighting male Rhinoceros beetles (1) are 16 cm long. A Hunting wasp (2) has caught a fly. An Assassin bug (3) feeds on a caterpillar pierced by its sharp jaws. A Southern hawker dragonfly (4) lays her eggs on a waterlogged branch. Black garden ant workers (5) tend aphids. They protect them from enemies and in return "milk" them for their sweet honeydew secretion. The grasshopper *Oedipoda miniata* (6) is well camouflaged, but if it is disturbed it jumps (7), showing its bright wings, before settling and "disappearing" again. The Monarch butterfly (8) makes long distance migrations from North America, where it feeds on the milkweed plant, south to Mexico where it hibernates. In the spring it migrates north again.

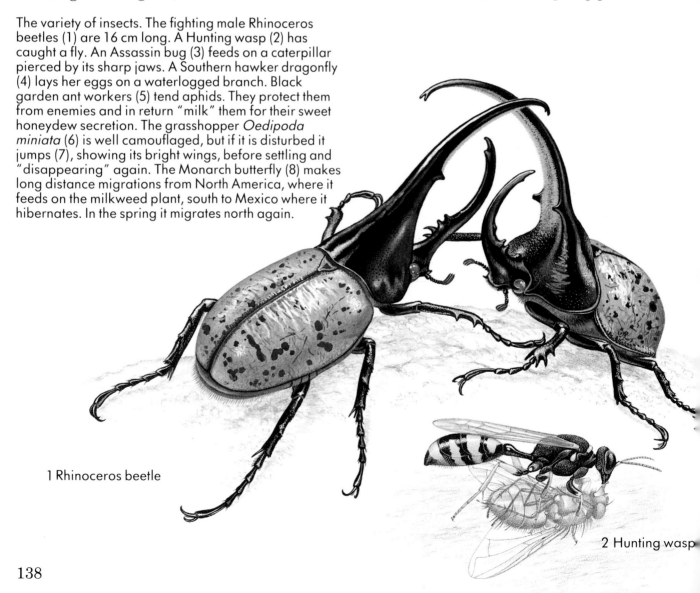

1 Rhinoceros beetle

2 Hunting wasp

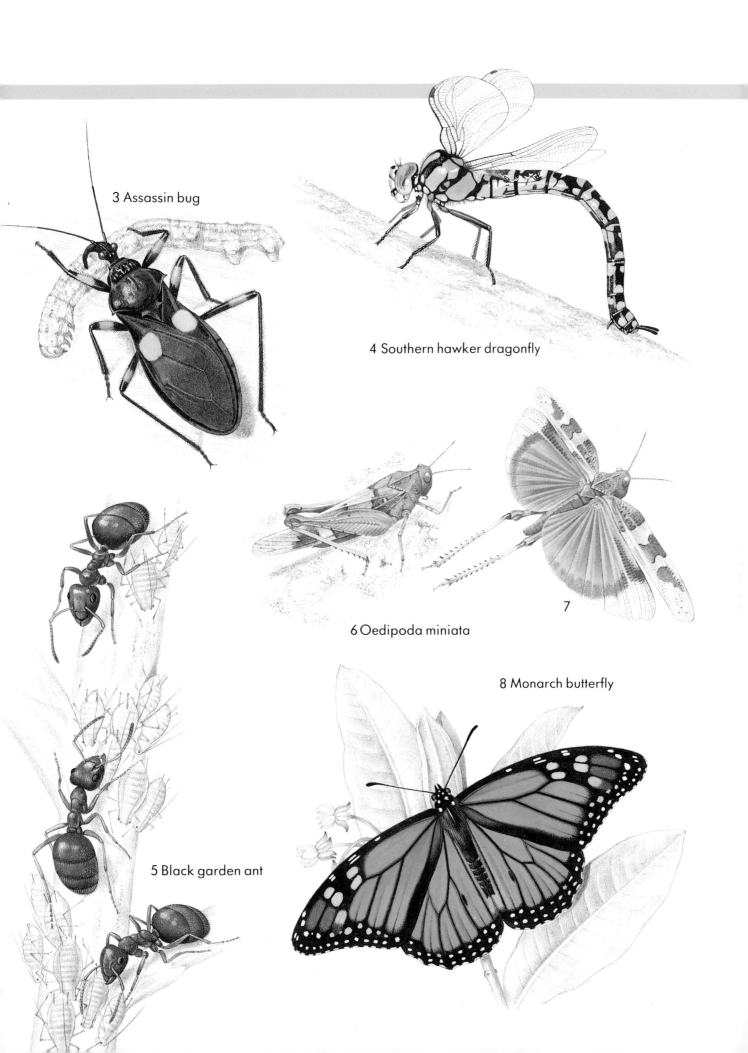

3 Assassin bug

4 Southern hawker dragonfly

6 Oedipoda miniata

7

8 Monarch butterfly

5 Black garden ant

Other jointed-legged animals

Apart from insects there are several other groups of jointed-legged animals, or arthropods. Centipedes get their name from their supposed 100 legs. In fact most have only about 40, although one species has 354. Centipedes are hunters. They have poison claws around the mouth. With these they catch insects and small worms. The largest tropical species, 33 cm long, can catch mice. Millipedes are a separate group. Instead of having one pair of legs on each segment like centipedes, they have two pairs. They are harmless vegetarians. Most feed on decaying plants. They live in damp, dark places or in the soil. Some tropical species are up to 30 cm long. Some have 750 legs, but none has 1,000, as their name would suggest.

Arachnids include spiders, mites and scorpions. Scorpions live in the warmer parts of the world. Like other arachnids they have four pairs of walking legs. In front of these they have large pincers which they use to seize and crush prey. They have a long tail with a sting at the end that curves over the back. This is used for defence and to subdue prey. Few species have a sting that is dangerous to humans. Scorpions mate after a "dance" in which the pair hold pincers. Scorpion young are born alive and climb on to the mother's back. There they stay until after their first moult. Like other arthropods they moult their outer skeleton as they grow and are vulnerable until the new larger coat has grown hard.

▼ Soldier crabs march across an Australian shore. Crabs are found on shores worldwide, and also in the depths of the sea. A few species in the tropics even live on land. Most crabs are hunters, or scavengers that feed on dead animals on the seabed. Some tropical species pick up pellets of mud and sieve them for food.

Spiders are all equipped with poison fangs to paralyse or kill prey. Not many species are dangerous to people. Most specialize in hunting or trapping insects. Spiders have special silk-producing glands. The silk they produce can be stronger than steel of the same thickness. The silk is used for safety ropes or parachuting. In many it is used for making webs to trap prey, or wrapping prey when caught. But some spiders do not use webs. Jumping spiders and wolf spiders pursue their prey on the ground. Crab spiders sit on flowers which match their colour, and ambush prey.

▼ A male orb-web spider courts a female by twanging her web (top). Female spiders are usually bigger than males, and males approach with care to avoid being mistaken for food. *Argiope bruennichi* (below) spins a strong sticky orb web to catch prey. Like many orb-web spiders it spins a new web each night.

Orb-web spiders

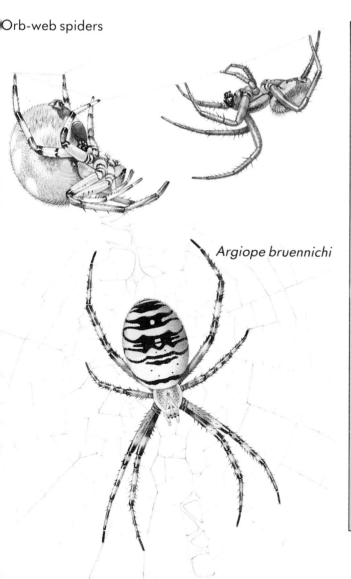

Argiope bruennichi

Most arthropods are land animals, although one large group, the crustaceans, has largely remained in the water. This group includes the crabs and lobsters and the many kinds of shrimp. There are almost 40,000 species. Crustaceans are important foods, both in fresh water and seawater, for many other animals. Some, such as the krill of the cold seas, may live in shoals of millions.

A lobster has a segmented body. The head has eyes and two pairs of antennae. The thorax and a long abdomen or tail follow behind. Each segment has a pair of legs. Those near the mouth work as jaws. Those under the thorax are walking legs. Those under the abdomen are swimming legs. A crab is much the same, but its small tail is tucked under the body. Other crustaceans show variations on this body plan. Some are quite extreme, such as the barnacle.

Spiny skins

A starfish tries to pull open a scallop shell to eat its flesh. Some starfish are fierce hunters. They belong, together with the sea urchins, in the phylum Echinodermata. These spiny-skinned animals are built on a five-rayed pattern. They have complicated bodies, with a skeleton of hard plates. They have a unique system of water canals which work the "tube feet" on which these animals move. The tube feet can also be used, as here, to grip prey. Sea urchins have the same basic body plan, but are round. Many eat plants, using their jaws to graze over rocks.

Fish, amphibians, reptiles

Nearly all big animals are vertebrates, or backboned animals. They are a very successful group. Their body chemistry works fast. Many are very active. They have developed body shapes and limbs to move in water, on land, and in the air. They have the biggest brains and include the most intelligent animals. Vertebrates first evolved in the water; later they colonized the land. Over hundreds of millions of years they developed more efficient bodies, brains, and ways of reproducing. But fish, amphibians and reptiles are still "cold-blooded" even though they have backbones.

► A Green python in an Australian rain forest. Snakes, like other reptiles and fish, amphibians, birds and mammals, have backbones. Some pythons have a backbone with more than 400 individual bones in it. These bones are called vertebrae.

Vertebrate ancestors

Fishes were the first of the vertebrates. We do not know all the stages by which they developed from simple invertebrates. Their immediate ancestors probably looked much like the living lancelet. Rather fish-shaped, the lancelet has blocks of muscle running down each side of its body. There is a tail fin, but no others. A mouth near the front lets in water which is passed out through gills. These obtain oxygen, and also filter food out of the water. This animal does not have a true backbone but has a stiffening rod, a notochord. Vertebrates have this during development before the backbone. The lancelet has no brain, skull or jaws, or true heart. The most primitive vertebrates, such a lampreys, still have no jaws. They have larvae that look very like lancelets. From beginnings such as this, true fish developed, with a proper backbone. They also developed fins, and heads with senses, a brain, skull and jaws.

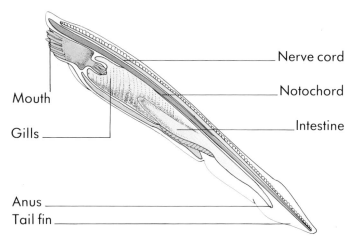

Mouth

Gills

Anus

Tail fin

Nerve cord

Notochord

Intestine

▲ The lancelet shows many of the features that might be expected in an ancestor of the vertebrates. Similar animals lived 550 million years ago.

▼ The red sea squirts have tadpole-shaped larvae with notochords, suggesting they are related to vertebrates.

Fish

There are two main groups of fish. The cartilaginous fish include the sharks and rays. They have rough skins with scales like little sharp teeth. In many ways they seem to be old-fashioned fish, and some have hardly changed for millions of years. But there are still over 600 species, among them the biggest fish of all, the plankton-eating Whale shark.

The bony fish are much more numerous. There are 20,000 kinds. They range from the large to the tiny. They are often streamlined to move fast through water, but there is an enormous variety of shapes and ways of life in this group. Eels, seahorses, catfish, flatfish, pike, puffer fish and flying fish are just a few of the shapes bony fish can take.

One small group of bony fish are called lobe-finned. They have paired fins with bones. They include the coelacanth and lungfishes. They are survivors of a group that was successful long ago. From their early relatives, all land vertebrates are descended.

Most bony fish belong to the group called ray-finned fish. Their paired fins are supported by thin rays. They breathe through gills. One gill cover covers all gills on one side. There is a single slit behind this to let out water that has come from the mouth via the gills.

Fish eat a huge variety of foods. Some strain the water for plankton. Others eat plants or snails. Some, such as piranhas, are fierce hunters with sharp teeth to match. Bony fish may have teeth not just on their jaws, but on the tongue, palate or throat lining too. Most bony fish lay large numbers of eggs which are fertilized externally and hatch as tiny larvae. Only a few kinds care for their young.

▶ A Great white or Man-eater shark homes in on a bait. The teeth work like a saw blade. Sharks have stiff pelvic fins that act like wings to give them "lift" in the water, as they have no swim bladder. They are not good at sudden turns or braking. The tail is not symmetrical. Its shape helps to lift the rear end as well as driving the fish forward. The mouth is below the head, and the large snout holds a good sense of smell. There is a row of separate gill slits on each side of the neck. The backbone is made of cartilage, or gristle, and not bone.

▼ A male sunfish stands guard with fins erect. The spines in the front part of the dorsal fin can be raised or lowered. The pelvic fins are close under the pectorals, giving good turning and braking ability. The swim bladder keeps the fish buoyed up. Bony fish such as this manoeuvre with ease.

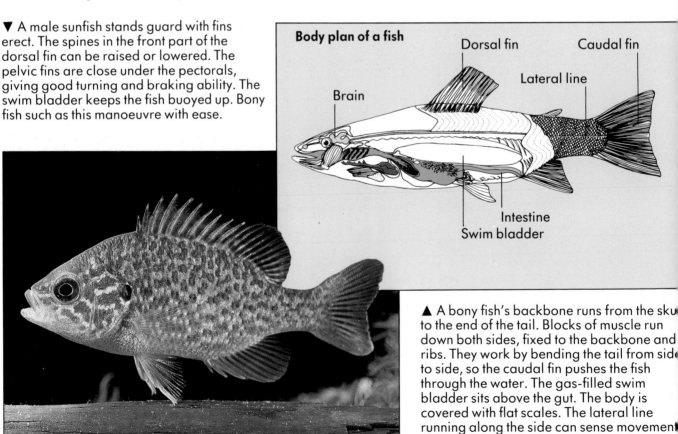

Body plan of a fish

Brain · Dorsal fin · Caudal fin · Lateral line · Intestine · Swim bladder

▲ A bony fish's backbone runs from the skull to the end of the tail. Blocks of muscle run down both sides, fixed to the backbone and ribs. They work by bending the tail from side to side, so the caudal fin pushes the fish through the water. The gas-filled swim bladder sits above the gut. The body is covered with flat scales. The lateral line running along the side can sense movement in the water. The fish breathes by taking oxygen from the water using the gills.

Amphibians

Amphibians include frogs, toads, newts and salamanders. They are mostly land animals, but their lives are tied to the water. They lay their eggs in water and these develop into tadpoles. These swim and have gills to breathe. Later they develop legs, lose their gills and develop lungs. At this stage, as tiny frogs or newts, they move to the land. Even then they mostly stay in damp places, because their smooth slimy skins are not waterproof. The advantage of these skins is that they can be used for breathing and help out the simple lungs.

Over 350 million years ago some lobe-finned fish lived where pools sometimes dried out. They had to wriggle to the next pool to survive.

Gradually some became better adapted for making these trips. The fins turned into legs. Air-breathing became important. Amphibians had arrived. The head of amphibians differs from that of fish. Amphibians have eyes with eyelids and tear glands, and blink to clean the eye surface. Fish have internal ears and hear well. On land something more is needed to pick up sounds in air. Amphibians have eardrums at the back of the head. For frogs, sound is important for attracting a mate.

Newts probably look most like the early amphibians. Frogs and toads are relatively new, but they are very successful, especially in the tropics. There are 3,500 species, compared with about 500 other amphibians.

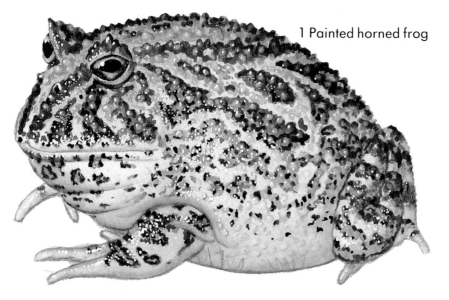

1 Painted horned frog

Varied amphibians. The Painted horned frog (1) of South America has a huge gape and eats mice and other frogs. A male Asiatic painted frog (2) calling. The Red salamander (3) of North America often burrows in mud. A Tiger salamander (4). A South African bullfrog (5) eats a rat. In this species, it is the males which guard the eggs and young. A male Koikoi poison-arrow frog (6) of South America carries developing tadpoles. A young Red-spotted newt (7). A Leaf frog (8) from the South American rain forest has toes with suckers to help it climb. A lungless salamander (9) from Central America is so tiny it can crawl about on leaves.

2 Asiatic painted frog

3 Red salamander

4 Tiger salamander

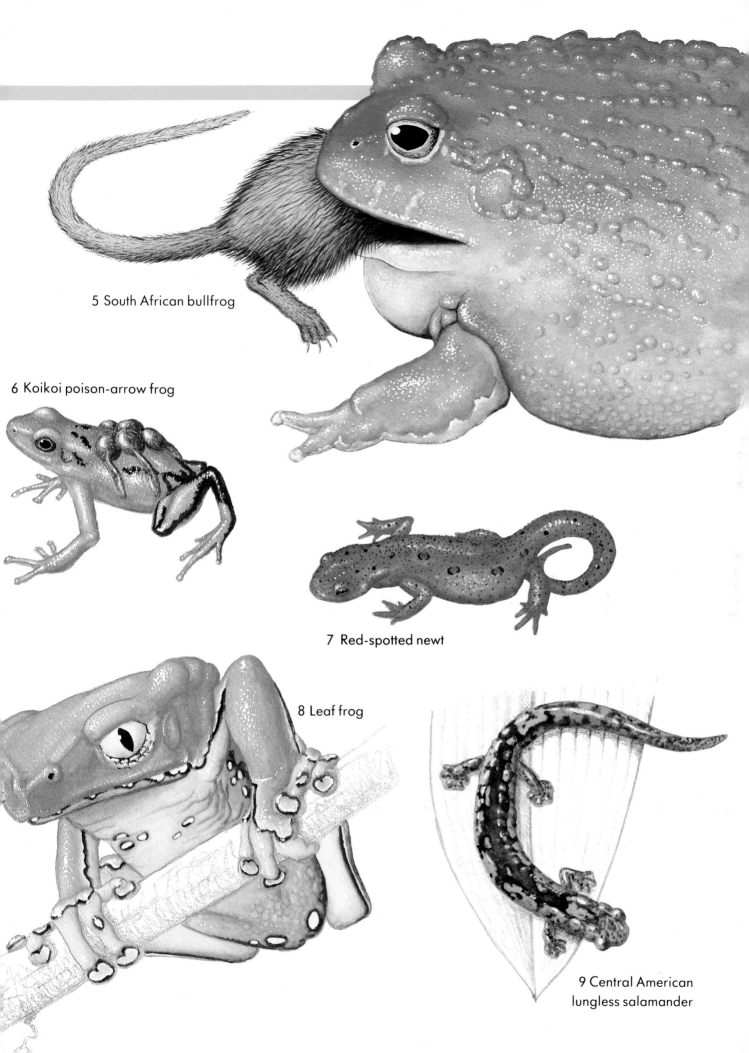

5 South African bullfrog

6 Koikoi poison-arrow frog

7 Red-spotted newt

8 Leaf frog

9 Central American
lungless salamander

Reptiles

Tortoises, crocodiles, snakes and lizards are all reptiles. Reptiles evolved from amphibians and in several ways are better at the job of living on land. Among other things their legs are generally longer and stronger, and their lungs are more efficient. But perhaps the two most important improvements reptiles have made are in their skins and in their breeding. Reptile skins are dry, scaly, and quite waterproof. Reptiles do not need damp places. Many live in deserts. They do not need water to breed either. Unlike the soft amphibian egg that has to develop in water, a reptile egg has a hard or leathery shell, and is laid on land. From this hatches out the young, which is a small replica of its parents.

Most reptiles do not look after eggs and young. Eggs are laid and left to hatch on their own. A reptile cannot incubate its eggs, as it is "cold-blooded". This means it depends on its surroundings for warmth, rather than generating heat in its body. Reptiles are most common in the tropics. Many bask in the sun to raise their temperature for activity.

Turtles are one of the most ancient reptile types, but also perhaps the oddest. The shell roughly corresponds to the ribs and scales of other reptiles. Land tortoises are usually slow moving plant-eaters. Terrapins and turtles are good swimmers and are mostly hunters. They must surface to breathe. Crocodiles also swim well, and are the largest living reptiles and fierce hunters.

Snakes and lizards are the most numerous reptiles. There are about 3,000 kinds of each compared to 250 species of tortoise and 25 of crocodile. Most lizards are small and feed on insects. They are found in every habitat from the tops of trees to burrows underground. Snakes are hunters and swallow prey whole.

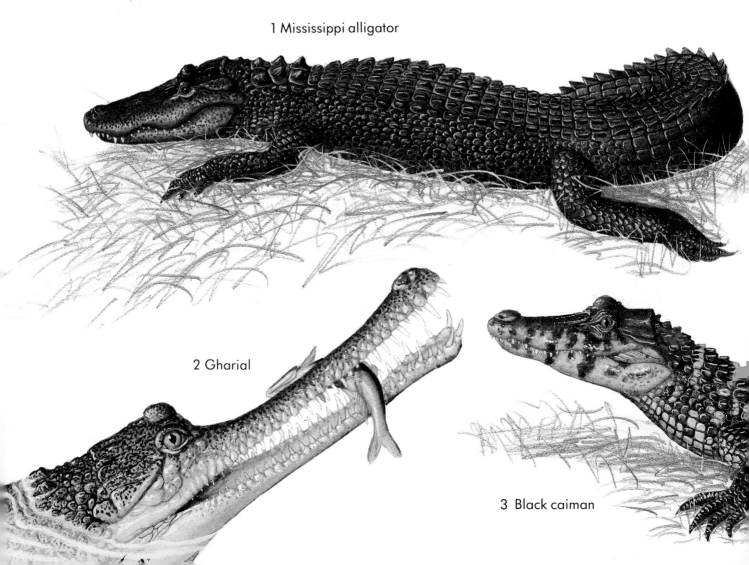

1 Mississippi alligator

2 Gharial

3 Black caiman

4 Flapnecked chameleon

5 Gopher tortoises

6 Green turtle

7 Cuban Island ground boa

A selection of reptiles. A Mississippi alligator (1) sits on her nest mound. Several kinds of crocodile are known to guard nests and young. The Gharial (2) is a narrow-snouted fish-eating crocodile from India. It grows to over 6 m long. The Black caiman (3) lives in South America. A Flap-necked chameleon (4) catches a fly. Chameleons can shoot out their tongues in a fraction of a second. These have sticky grasping tips. Gopher tortoises (5) are good burrowers. These are mating. A Green turtle (6) feeds on sea grasses. The Cuban Island ground boa (7) is one of the smaller constricting snakes. It swallows prey whole, and digests it slowly.

Birds and mammals

Spot facts

● *A hummingbird's wings may beat up to 80 times a second.*

● *The longest time a mammal spends growing in its mother before birth is the 21 months of the Indian elephant.*

● *In flight, the heart of a small bat beats 1,000 times a minute.*

● *At up to 33 m long, the Blue whale is the biggest living animal.*

● *The kiwi of New Zealand lays the largest egg for its size of any bird. It may be one fifth the weight of the mother.*

▶ A baby Common brushtail possum of Australia attached to a teat in its mother's pouch. At birth it weighed only 0.2 g. One of the characteristics of mammals and birds is their care of their young. This gives the babies a good start in life. Only small numbers of young need to be produced, in order for some to survive.

Birds and mammals are the most highly evolved vertebrates. They are "warm-blooded" and able to keep themselves at a constant temperature in a wide variety of surroundings. So they can always be active, even in cold conditions which would stop a reptile's body working.

A disadvantage of this is that a warm-blooded animal needs more energy, and must feed regularly. Insulation is also needed on the outside to conserve heat. Birds and mammals have the largest brains among the vertebrates. Some are very intelligent. They are also very good at looking after their young.

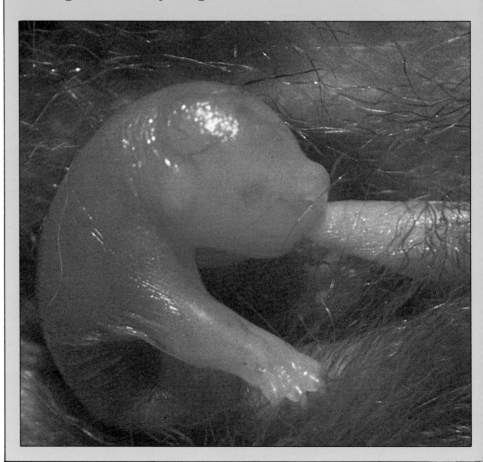

There are some 8,500 species of birds, and they live everywhere from the tropics to the Arctic and Antarctic. Birds have a body covering of feathers to keep them warm, but still have scales on their back legs like their reptile ancestors. Their front legs have become wings, but still show the same basic arrangement of bones as a reptile.

Birds lay hard-shelled eggs, then sit on them to keep them warm until they hatch. Some lay eggs in little more than a scrape on the ground. Others build elaborate nests for the eggs and young. After hatching, some young are able to leave the nest immediately. They stay with their parents for a while for protection and to learn to find food. Other birds hatch blind and naked, and are kept warm and fed by the parents in the nest before fledging.

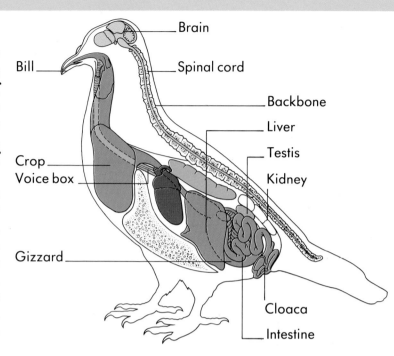

▼ A flock of Galah cockatoos fly through the Australian countryside. They are a pest on crops, and their flying ability means they can travel considerable distances to find food. The birds' power of flight is also useful to help the birds escape from their enemies.

▲ Anatomy of a bird. The heart is large, as might be expected in a very active animal. The brain is well developed. The backbone runs from the back of the skull to the base of the tail. A bird has no teeth. The gizzard helps to grind up the food which is swallowed.

Birds

Considering the advantages of flight it is surprising how many groups of birds have become flightless. Ostriches, cassowaries and rheas are all good runners, but cannot fly. Penguins' wings have become adapted for "flying" through the water rather than air. The biggest flying birds weigh about 15 kg. They include the Mute swan, the Great bustard and the Andean condor. The Wandering albatross, which spends most of its life airborne gliding across the oceans, has, at 3.5 m, the longest wingspan.

At the other extreme is the tiny Bee hummingbird which weighs 1.6 g. Hummingbirds can hover in one spot to feed on nectar. Other birds including sunbirds and some small parrots like this form of food. A huge number of birds are insect-eaters. They may have slender bills to pick insects off leaves. Some have other techniques, for example swifts and swallows which pursue insects through the air using the beak as a net. Seed-eaters include finches with triangular bills to match the size of seed they eat, pigeons that swallow grain whole, and parrots with powerful pincer beaks. The biggest parrot, the Hyacinthine macaw, can easily crack Brazil nuts. Other birds kill large prey, or eat carrion or fish.

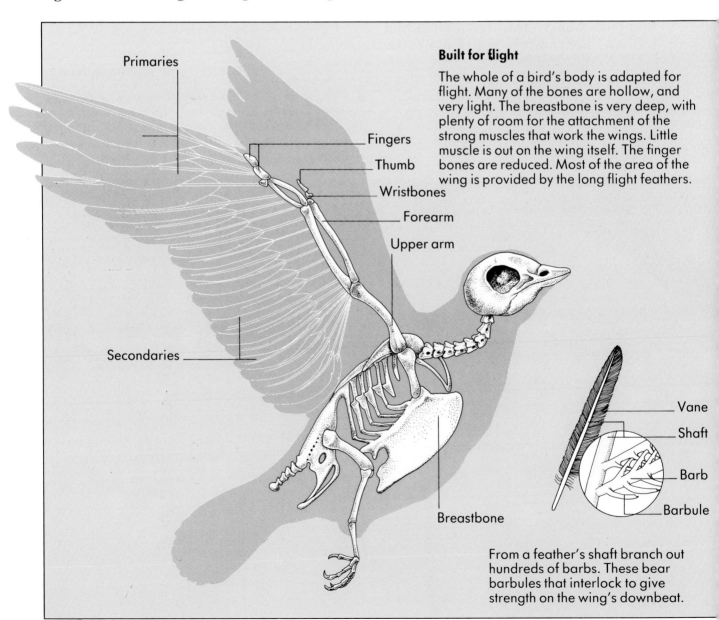

Primaries

Fingers

Thumb

Wristbones

Forearm

Upper arm

Secondaries

Breastbone

Built for flight

The whole of a bird's body is adapted for flight. Many of the bones are hollow, and very light. The breastbone is very deep, with plenty of room for the attachment of the strong muscles that work the wings. Little muscle is out on the wing itself. The finger bones are reduced. Most of the area of the wing is provided by the long flight feathers.

Vane

Shaft

Barb

Barbule

From a feather's shaft branch out hundreds of barbs. These bear barbules that interlock to give strength on the wing's downbeat.

Contrasting birds. The Tufted duck (1) is a good diver, paddling with its feet in pursuit of small fish and insects. The Whiskered tern (2) has a buoyant flight. It hovers above the water then darts down when it spots a small fish. The Marsh harrier (3) is a bird of prey. It has a sharp hooked beak, and sharp claws for catching prey. The ostrich (4) is the largest bird and is completely flightless, although it has quite large wings, which it uses in display. The legs are long and end in two hoof-like toes. It has the biggest eyes among the vertebrates. The New Zealand South Island brown kiwi (5) is also flightless and has tiny wings. It is active at night and probes the ground for worms using its sensitive beak.

2 Whiskered tern

1 Tufted duck

3 Marsh harrier

4 Ostrich

5 South Island brown kiwi

153

Classifying the mammals

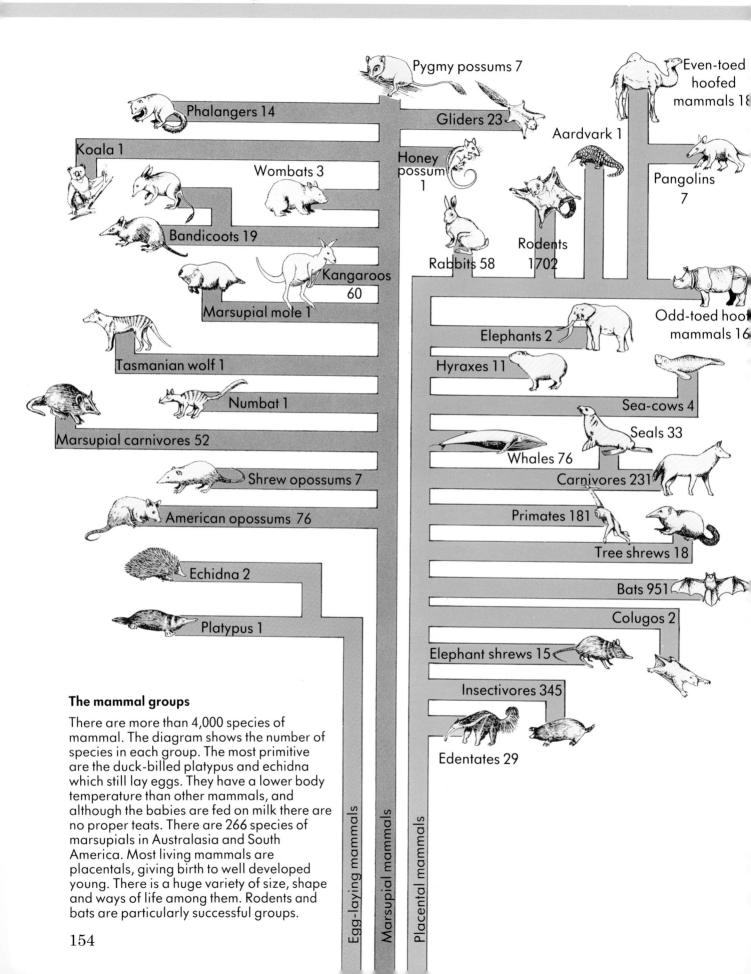

Pygmy possums 7

Phalangers 14

Gliders 23

Even-toed hoofed mammals 18

Aardvark 1

Koala 1

Honey possum 1

Pangolins 7

Wombats 3

Bandicoots 19

Rabbits 58

Rodents 1702

Kangaroos 60

Marsupial mole 1

Odd-toed hoofed mammals 16

Elephants 2

Tasmanian wolf 1

Hyraxes 11

Sea-cows 4

Numbat 1

Whales 76

Seals 33

Marsupial carnivores 52

Carnivores 231

Shrew opossums 7

Primates 181

American opossums 76

Tree shrews 18

Bats 951

Echidna 2

Colugos 2

Platypus 1

Elephant shrews 15

Insectivores 345

Edentates 29

Egg-laying mammals

Marsupial mammals

Placental mammals

The mammal groups

There are more than 4,000 species of mammal. The diagram shows the number of species in each group. The most primitive are the duck-billed platypus and echidna which still lay eggs. They have a lower body temperature than other mammals, and although the babies are fed on milk there are no proper teats. There are 266 species of marsupials in Australasia and South America. Most living mammals are placentals, giving birth to well developed young. There is a huge variety of size, shape and ways of life among them. Rodents and bats are particularly successful groups.

Mammals keep a constant body temperature. Most have a coat of fur to help stay warm. With a thick coat a polar bear or Arctic fox can live in the far north with temperatures at minus 40°. Mammals can live equally well in the tropics. Although the largest may not need fur for warmth, even elephants have some hair on the body. Fur can be raised or lowered to help adjust temperature. Shivering, sweating and panting can also play their part.

Placental mammals have babies born after a relatively long gestation period. Some, such as antelopes, may be able to walk soon after birth. Others, such as mice or dogs, are quite helpless when they are born. In both cases the babies are fed on milk by the mother. Even when they are weaned on to solids many mammal young stay with their mother. They are able to learn while still getting protection. In some species it may be years before they are totally independent.

Placental mammals probably originated from ancestors that looked like shrews. From these beginnings the shapes of mammals' bodies have changed to produce good runners. There are also mammals which are good climbers, swimmers, burrowers and fliers. Body size ranges from that of the 150-tonne Blue whale to that of a bat which weighs only 1.5 g.

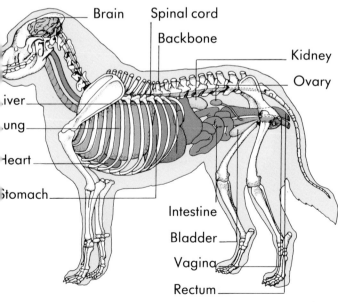

The anatomy of a dog. Like most placental mammals it has four well-developed limbs, large lungs and an efficient heart. The legs are below the body, supporting the weight easily. The brain is large. This is an animal suited to a high degree of activity.

Marsupials

An Eastern grey kangaroo of Australia and pouch young. Like other marsupials it is born after a very short gestation period and completes its development in the mother's pouch. The baby is smaller than your thumb when it is born. It is blind and has large front legs and small back legs. It crawls to the mother's pouch and there fastens on a teat. It stays in the pouch for months until it is kangaroo-shaped and able to move outside. Marsupials have developed many different lifestyles. Kangaroos themselves are like antelopes in the way they deal with plant food. Their hopping gait is unusual, but can be fast. A large kangaroo may cover 9 m in one bound. There are marsupial moles, anteaters, hunters, burrowers and gliders that parallel mammals in other parts of the world. All have babies born as little more than embryos, which fix on to a teat. But not all marsupials have a proper pouch.

155

Carnivores

Mammals of several different groups, or orders, are carnivores in the sense of being flesh-eaters. But one particular order is known as the Carnivora, because nearly all its members specialize in a meat diet. These carnivores include the dogs, cats, bears, raccoons, weasels and mongooses.

These animals have long, pointed canine teeth for biting to kill. The cheek teeth may be few in number, but have pointed ridges so that the top and bottom jaw can act together as scissors for slicing meat. Hunters need good eyes, ears and nose for finding prey. Then they need speed to catch it. Many of the carnivores, such as dogs and cats, stand on their toes rather than the soles of the feet. Agility and being able to squeeze through burrows may also be useful, as is seen in some weasels and mongooses.

Meat can be quick to digest and nutritious. I can be roughly chopped and swallowed in lumps. Only a short digestive system is needed Once it has had a good meal, a carnivore may not become hungry again for some while. In between hunts, it can afford to be lazy. A lion may have a good meal once or twice a week. It may sleep for 20 hours a day. It needs to do little until it is hungry again.

Other hunters include insect-eaters such as shrews and bats, with small pointed teeth. Those that eat ants, such as pangolins, may do without teeth, but have long sticky tongues to pull in their meal. Fish-eaters such as dolphins and some seals have pointed teeth to catch the slippery prey. They are streamlined for pursuing prey through the water and their limbs have been modified into flippers.

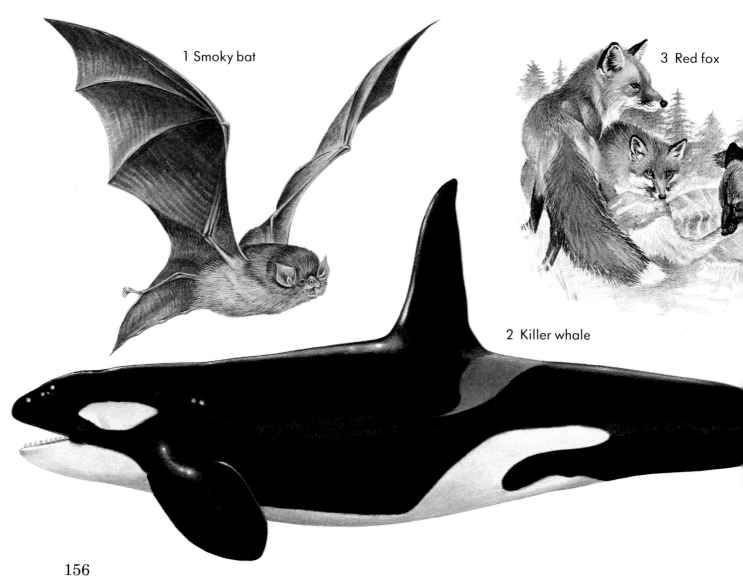

1 Smoky bat

3 Red fox

2 Killer whale

Animals that prey on others. The Smoky bat (1) of South America feeds on small insects caught on the wing. The Killer whale (2) grows 8 m long and can swim at 50 km/h. It is one of the fiercest hunters in the sea. It catches seals and other whales. The Red fox (3) kills small animals such as rabbits and voles, and may also scavenge, like the four shown here. It also eats worms. Like some other mammals, this species can vary in colour. The American black bear (4) may attack prey as large as deer, but is comparatively slow-moving. It also feeds on berries and other plants. The Fat-tailed dunnart (5) is a mouse-sized marsupial that is a fierce hunter of insects. The Algerian hedgehog (6) eats insects and any other small animals it can catch. The False killer whale (7) catches squid and large fish with its sharp teeth. The South American sea lion (8) dives in pursuit of squid and large shrimps.

4 American black bear

5 Fat-tailed dunnart

6 Algerian hedgehog

7 False killer whale

8 South American sea lion

Herbivores

Several groups of mammals are adapted as herbivores, or plant-eaters. The rodents, such as mice, squirrels and beavers, have large teeth for gnawing at the front of their jaws, and a battery of chewing, grinding teeth along the side. As well as eating grass or leaves, many are able to gnaw into tough foods like bark or nuts. Most rodents are small animals. Rabbits and hares are not rodents, but they have the same types of teeth and feed in similar ways.

Most of the big plant-eaters are hoofed mammals. There are two groups. The odd-toed hoofed mammals are horses, tapirs and rhinos. Among the even-toed group are deer, camels, antelopes and cattle. Some of these animals browse leaves and twigs from bushes. Giraffes reach high into the trees with tongue and lips. Other hoofed animals have broad lips to feed from the ground, munching tough grasses.

Many plants have little food value, so plant-eaters have to eat a lot. They spend most of the day feeding. They must always be alert for enemies. The food needs a good chewing by the grinding teeth. The food is hard to digest, and so the gut is long. Even-toed hoofed animals have complex stomachs where food ferments. They bring balls of food, the cud, back to the mouth for a second chewing. Odd-toed hoofed mammals do not chew the cud, but bacteria in the intestines help digestion.

1 Topi antelope

2 Plains zebra

3 European hare

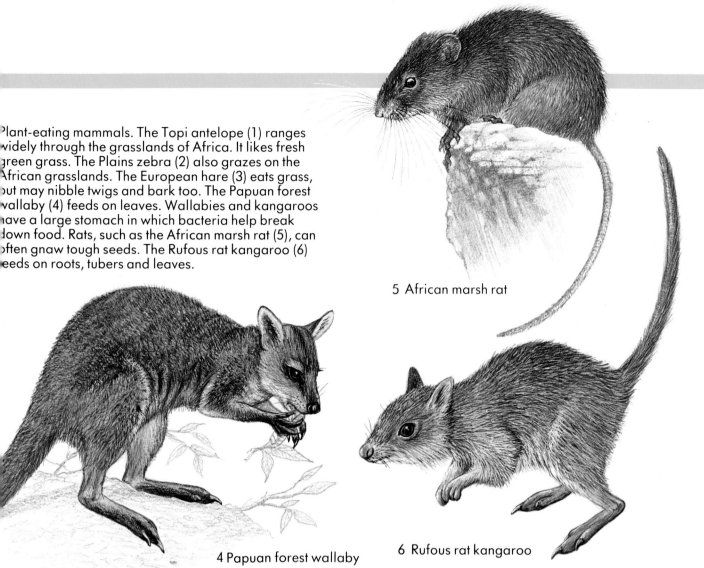

Plant-eating mammals. The Topi antelope (1) ranges widely through the grasslands of Africa. It likes fresh green grass. The Plains zebra (2) also grazes on the African grasslands. The European hare (3) eats grass, but may nibble twigs and bark too. The Papuan forest wallaby (4) feeds on leaves. Wallabies and kangaroos have a large stomach in which bacteria help break down food. Rats, such as the African marsh rat (5), can often gnaw tough seeds. The Rufous rat kangaroo (6) feeds on roots, tubers and leaves.

5 African marsh rat

4 Papuan forest wallaby

6 Rufous rat kangaroo

Primates

The order of mammals we belong to is known as the Primates. It includes monkeys and apes. Most primates live in the trees. They have eyes that face forward to judge distances as they jump and climb. They have good grasping fingers and toes, and have pads on the end of the fingers and nails rather than claws. Some South American monkeys, such as the Brown capuchin, can grasp with their tail too. Many monkeys feed on both animal and plant food. The adaptations monkeys have for climbing also make them good at investigating their surroundings. They can look carefully, and use their hands to touch objects, or move or pull them apart. These adaptations were also important to our ancestors, allowing them to develop their intelligence.

Brown capuchin

Sooty mangabey

Night monkey

159

Feeding the world

- *About 85 per cent of the world's food comes from just 20 different plant species.*

- *Agriculture accounts for more than 70 per cent of the world's annual water consumption in the form of irrigation.*

- *On average between sixty and eighty thousand people starve to death each day.*

- *The United States uses over 125 kg of artificial fertilizer per hectare of cultivated land every year.*

- *Britain has more farm tractors (300,000) than farm workers (250,000).*

- *Women grow over half of the world's food. In Africa, they grow 75 per cent of it.*

► Mothers and children in Ethiopia eating the daily rations provided by famine relief agencies. Today, the world produces enough food to feed an additional 1,000 million people, but people still die from hunger. The problems of world hunger are largely the result of inefficient distribution.

Nearly all of the world's food comes from agriculture, the deliberate cultivation of plant and animal species. Agriculture exploits the natural processes of growth and reproduction, but people have increasingly altered the course of nature in order to feed our unnaturally large population.

In general, the development of agricultural techniques has been a gradual process. Much of the world's food is still grown using methods that have not altered in centuries. But during the last 40 years global food production has doubled. Most of the increase has been due to the application of modern scientific techniques, such as selective breeding, and the introduction of machinery, artificial fertilizers and pesticides.

Early days

▲ A 3,000 year old Egyptian wall painting. The artist has depicted typical farming activities, such as ploughing and harvesting, together with a wide range of crops, including grains, fruit and vegetables.

Civilization requires a settled existence. Towns and cities depend upon a surplus from the fields to feed populations that are not growing their own food. The development of civilized skills, such as reading, writing and painting, in turn depends upon urban populations freed from daily labour in the fields.

The most primitive form of human lifestyle is that of the hunter-gatherers. Wild animals are trapped and hunted for meat and skins, but fruits, nuts and seeds gathered from the wild provide the basis of daily existence. Hunter-gatherers are always in danger of outgrowing their food supply. If there are too many mouths to feed, the available food resources can quickly be used up.

Agriculture developed from attempts to improve this precarious lifestyle, and probably began with the domestication of the first food animals. Wild sheep and goats were collected into herds and managed, rather than being hunted and killed.

Animal herders have a nomadic existence; they must be constantly on the move in search of fresh pasture. Long-term human settlement, such as the first permanent towns, only became possible when we discovered how to domesticate and cultivate certain nutritious plants.

The earliest evidence of cultivation comes from the Middle East around 8,000 years ago. About that time, the ancestors of wheat and barley were first sown in fields that were worked entirely by hand. In Mexico, maize (corn) was grown about 6,500 years ago, and in China rice was first cultivated around 5,000 years ago.

In ancient Egypt, irrigation water provided by the annual flooding of the River Nile enabled farmers to produce large surpluses of food.

161

Ancient and modern

The two most primitive forms of agriculture still in existence are nomadic herding and tropical slash-and-burn cultivation. Neither requires any specialized technology, but neither is capable of producing a large food surplus. Increased productivity only became possible through the introduction of new technology such as the plough.

The ox-drawn plough was in use before 2500 BC, and the ox remained the main work animal until about AD 1000 when the horse collar was invented. Horsepower became dominant in Western agriculture, although in other regions, oxen and cattle are still used. In many parts of Asia, Africa and South America, the hoe remains the main agricultural implement. In the developed world, the introduction of machinery and the invention of the tractor and combine harvester, greatly improved agricultural output.

The ability to produce a surplus frees farmers from "subsistence agriculture", growing just enough to feed the family. Instead, they may grow a little extra to sell. This is a "cash crop". Today, about half the world's food, and many other substances, are produced as cash crops by specialist farmers. The remainder of our food is produced by millions of subsistence farmers.

▲ Sudanese tribesmen threshing sorghum, a coarse grain. Primitive agriculture requires no tools other than a digging stick, which doubles as a threshing club. Often the crops receive little or no attention between planting and harvest. Sorghum and millet are the main cereal crops in dry countries.

◀ Planting crops in Brazil, on land cleared by slash-and-burn, an effective technique for forcing a good crop from poor forest soils. Burning the trees releases nutrients, in the form of wood ash, back into the soil. Growing crops make good use of these but the soil is left exhausted. After one harvest, the land must be left at least 30 years for the forest to re-grow in order to recover its fertility. Slash-and-burn is also practised in tropical Africa and South-east Asia.

Efficiency in agriculture is a very difficult concept to define. At the most basic level, it is the amount of food (the yield) that can be obtained from a certain area of land. In this sense, the new hybrid crops are much more efficient than any of the traditional varieties. A more precise definition of efficiency also needs to take into account the costs of producing the crop in terms of irrigation, fertilizer and labour.

The type of crop is also important. Soya beans, for example, are about 10 times more efficient at producing protein than cattle. But Western tastes still demand meat, and much of the soya bean crop goes to feed the cattle. They are a more desirable food product, despite the expensive use of land. Western farming is considered the most efficient type of agriculture, but there are some serious disadvantages. For example, pollution from agricultural chemicals has now entered water supplies, which may be dangerous.

▲ Aerial spraying is often used to spread artificial fertilizer and pesticide over large fields. The technique wastes a great deal of chemicals, but is considered efficient because it saves time and labour costs.

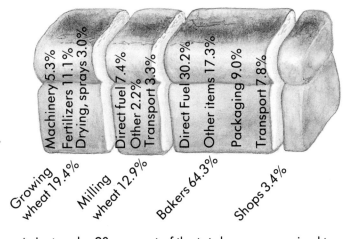

Growing wheat 19.4%
Machinery 5.3%
Fertilizers 11.1%
Drying, sprays 3.0%

Milling wheat 12.9%
Direct fuel 7.4%
Other 2.2%
Transport 3.3%

Bakers 64.3%
Direct Fuel 30.2%
Other items 17.3%
Packaging 9.0%
Transport 7.8%

Shops 3.4%

▲ Just under 20 per cent of the total energy required to produce a loaf of white bread is consumed in growing the wheat. Over 60 per cent is accounted for by the baking, packaging and transport.

163

Selective breeding

Selective breeding is as old as domestication itself. Humans have influenced the evolution of other species by tending only those plants and animals with desirable characteristics. Weak and diseased individuals have been removed from domesticated populations.

By 1930, the breeds and varieties used in agriculture were the result of centuries of luck, folk-wisdom and fashion. During the last 50 years, the application of genetic science has given us the power to shape plants and animals more precisely to fit our needs.

A worldwide breeding programme, involving researchers as far apart as Chile and Japan, has produced a series of improved cereal varieties. Some have provided an immediate 20 per cent increase in yield; others ripen during shorter summers, or resist certain plant diseases. These hybrid varieties now dominate world production. In the case of wheat, average yield has increased threefold. Other crops have received similar attention, and scientists are currently trying to introduce insect-repellent genes from a wild potato into ordinary garden varieties.

The demands of the meat-eating populations in Europe and America slowly changed during the last century. Animal fat was no longer needed to make candles, and increased vegetable oil production meant that less fat was needed as food. What people wanted was lean meat.

Animal breeding programmes have pioneered many modern techniques. The first test-tube babies were in fact calves. The widespread use of artificial insemination has allowed a rapid transformation of livestock populations. Modern beef cattle have a higher ratio of muscle to bone, and our dairy herds produce more milk. Our pigs are leaner, and our sheep produce more lambs and wool. But they lead unnatural lives, supported by the increasing use of drugs.

These new techniques have played a major role in boosting animal productivity to its present limits. Further changes will be possible with the application of genetic engineering to food animals.

The Green Revolution

The new hybrid crops produced bumper harvests in Europe and the United States during the 1950s. During the 1960s they were introduced into many developing countries. By 1979 more than 50 million hectares had been planted with hybrid seeds. The result was a Green Revolution that greatly increased productivity without extra land being put under cultivation.

In India, cereal production doubled in less than 20 years; in the Philippines the rice yield went up by 75 per cent. As well as producing more grain per hectare, the high-yield varieties mature more quickly. Some rice farmers are now able to achieve three crops per year.

Between 1950 and 1990, the Earth's human population doubled, and so did agricultural output. The Green Revolution has enabled the Earth to continue to feed the human population, but at a price.

Many of the new varieties will only perform well if they receive the artificial fertilizers that are commonplace in developed countries. These must be paid for, often as expensive imports, along with pesticides and other additives. The widespread use of agrochemicals in developing countries is now a significant factor contributing to increased global pollution.

Triticale (right), a hybrid of wheat and rye, was developed during the 1950s, and was the first completely new crop for thousands of years. It is now a heavy-yield grain crop in many countries.

▲ Hybrid maize (corn) growing in a genetic engineering laboratory in California. Genetic engineering allows scientists to alter a plant species without crossbreeding thousands of individual plants.

Santa Gertrudis

Kankrej

▲ The Santa Gertrudis bull is of a modern strain, bred from zebu cattle like the Kankrej . Both do well in hot climates, but the Santa Gertrudis is farmed for beef on a large scale in Texas, whereas the Kankrej is a draft animal in India.

165

Crop farming

● *On average, our crops require 1,000 kg of water to produce 1 kg of food.*

● *Plants only use about 1 per cent of the energy available in sunlight when making food by photosynthesis.*

● *Half the world's food comes from just three plants – wheat, rice and maize (corn).*

● *A European farmer can produce enough wheat on a single hectare to feed someone on a European diet for one year.*

● *The world's largest single field is used for growing wheat, and covers 14,160 hectares of the Canadian province of Alberta.*

▶ Terraced rice paddies in the Philippines. Terracing is a useful way of conserving valuable water, and making more land available for farming. Crops such as vines and olive trees are also grown on terraced hillsides in other parts of the world. Water can also be conserved on flat land by building low stone walls around crops.

The crops in our fields are the plant species that we have selected for cultivation. Many crops are restricted by climate to those parts of the world where they originated. Others have been transported around the world by human colonists.

The most important crops are known as staples and form the bulk of daily food intake. The staple crop varies in different parts of the world, but in most places it is one of the crops known as cereals.

Fruits and vegetables are usually grown to add variety to our diet, although some, such as potatoes and pulses, are also highly nutritious foods. Other crops are grown to provide basic ingredients, luxury goods such as tea and coffee, or raw materials such as vegetable fibres.

The farming year

In all parts of the world, the rhythm of farming is dictated by the growing cycle of the main crop. Most of our crops are annual plants. Harvesting food often means taking the seeds, and these crops have to be sown afresh each year. Some, notably rice, have a very short growth period, and in the right climate can be sown more than once a year. Others, like fruit trees, supply an annual harvest for years.

Farming removes much greater amounts of water and nutrients from the soil than natural vegetation. In many countries, the supply of additional water, through irrigation, is an essential feature of agriculture. Rice cultivation depends on irrigation by flooding, but for other crops, such as maize, spraying is the most common method although it is extremely wasteful of water.

Traditionally, nutrients have been replaced by applying natural fertilizers, or by letting the land rest and lie fallow for a year. The most common natural fertilizers are animal dung and compost from decaying vegetation. Letting land lie fallow slowly developed into systems of crop rotation. Some crops, such as clover, are useful only for animal grazing, but are very efficient at putting nutrients back into the soil.

The introduction of artificial fertilizers in the 1900s has transformed agriculture. Farmers can now grow the same crop on the same land year after year, without exhausting it. Chemical weedkillers have also made the farmer's task much easier, as has the increased use of machines. But crop farming is still hard work. The land must be prepared each year, and the crops tended at all stages of growth.

▼ Western crop farming is based on the growing cycle of the staple crop: in this case wheat. In the warmer regions, winter wheat can be sown in the autumn. Where the winter frosts are cold enough to kill the developing plants, spring wheat is sown, to be harvested in late summer. In the West, seeds are sown by a machine called a seed drill, and crops are harvested by a combine harvester. Crops are top-dressed with fertilizer and pesticide.

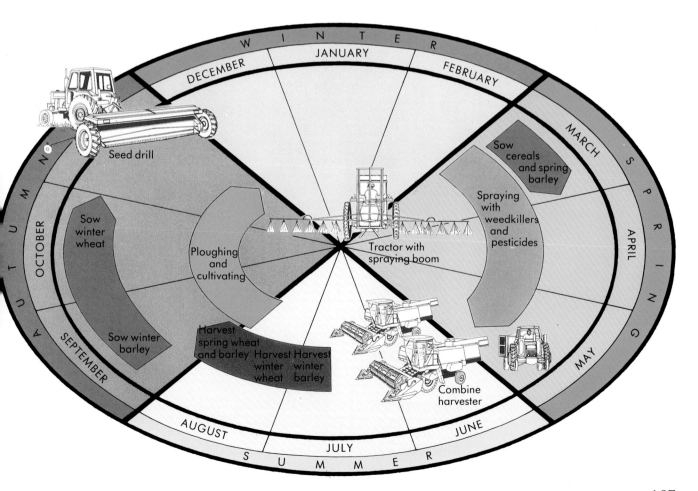

167

Cereals

Cereals, the grain producing plants, are our most important crops. The world produces about 1,500 million tonnes of cereals each year. Cereals are an ideal human foodstuff, mostly carbohydrate but containing 6 to 11 per cent protein. Without these essential crops we could not even begin to feed the world.

Wheat is the most common cereal. In the form of bread or pasta, cooked wheat is the staple food of more than a third of the world's population. Most of our wheat is grown in the temperate climates of Eurasia and America, but it is also widely cultivated in India, China and Australasia.

Wheat is most productive when farmed extensively in large open fields covering hundreds of hectares. Under these conditions, modern machinery, such as combine harvesters, can operate at maximum efficiency.

Wheat farming in the developed countries is now almost completely mechanized. In Africa and Asia, wheat is generally grown in much smaller fields that are sown and harvested by hand. Small farmers often lack machinery.

Rice is the staple cereal of tropical Asia, and requires a unique system of wet cultivation. Modern rice is not a seasonal crop, but the plant must be submerged in water for about 75 per cent of its growing period. Rice cultivation creates a distinctive landscape of flooded paddy fields. Mechanization is difficult, especially on terraces, because of potential damage to paddy walls and the submerged crop. The Water buffalo is far more important than the tractor and human labour most important of all.

Maize (corn) originated in South America where it is still grown by traditional methods. It is also grown as food in parts of Africa. In the developed countries it is cultivated using modern machinery, and is mainly used for animal feed.

Millet is the world's fourth most important cereal, and is a staple crop in the drier regions of Asia and Africa. Rye, oats and barley are grown in areas such as northern Europe, which have a cool damp climate. These three grains are little used as human foodstuffs, and nearly all the harvest goes as animal feed.

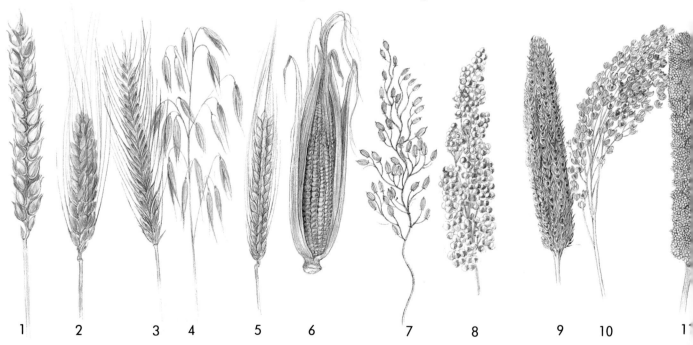

1 2 3 4 5 6   7 8 9 10 11

▲ The true cereals are all members of the grass family and are grown all over the world. Corn (maize) is most important in America, wheat in Europe, and rice in the tropics. The other cereals are of great importance in Africa.

Key
1 Bread wheat
2 Hard (Durum) wheat
3 Rye
4 Oats
5 Six-rowed barley
6 Maize
7 Rice
8 Sorghum
9 Finger millet
10 Common millet
11 Foxtail millet

▲ A rice harvest in Nepal. Some of the new varieties of rice are faster growing as well as giving a higher yield. Some farmers can now grow three harvests per year. Such results depend on "supporting" the crop with artificial fertilizers and pesticides, and many farmers cannot afford these expensive chemicals. Although more food is now grown on the same amount of land, on average it costs more to produce.

▲ Rice is one of the very few crops that requires constant irrigation. The cultivation of rice is mainly limited to river valleys with a year-round supply of water. Individual farms usually cover less than 20 hectares because of the amount of work involved. Each rice seedling must be planted by hand, and so fields remain limited to the size a family can work in a day. Fish and ducks are traditionally raised in paddies alongside the rice plants.

◀ Wheat is grown on every continent. Asia is the world's leading producer, but the North American prairies produce the greatest surplus. British wheat farming is the most productive in the world, with yields up to 15 tonnes per hectare.
   The huge machines used for intensive wheat production give rise to a characteristic landscape, with few obstructions. On hills, contour farming may help prevent soil erosion by rainwater.

Fruit

Fruit and vegetables provide variety, vitamins and roughage to diets that would otherwise consist of carbohydrate and protein. A few are grown as subsistence crops, and potatoes and cassava have become staples in some countries. Others, such as pineapples and bananas, are grown as cash crops, or like turnips as winter feed for livestock.

Different crops are cultivated for different parts of the plant. In general, the seeds and fruits are the most nutritious parts, but the leaves, stems and roots of some are also eaten.

Fruit is usually cultivated as a plantation crop. An orchard is just a less intensive form of plantation agriculture, and is more suited to temperate climates. Harvesting fruit is extremely labour-intensive, as great care must be taken not to damage the crop. For this reason fruit is usually an expensive luxury item. Tropical fruit is often picked before it is ripe so that it will remain fresh during transportation. It is shipped in refrigerated containers. Worldwide, an increasing amount of fruit is now grown for processing, especially into juice.

Some common tropical fruits

Some common temperate fruits

1 Pineapple 2 Durian 3 Carambola 4 Mango
5 Pawpaw 6 Soursop 7 Persimmon 8 Mangosteen
9 Pomegranate 10 Litchi 11 Akee 12 Cherimoya
13 Banana 14 Guava 15 Sapodilla 16 Passion fruit
17 Loquat 18 Cape gooseberry 19 Rambutan

1 Blackcurrant 2 Redcurrant 3 Bilberry 4 Blueberry
5 Sweet Cherry 6 Cranberry 7 Black mulberry
8 Gooseberry 9 Raspberry 10 Plum 11 Strawberry
12 Fig 13 Damson 14 Greengage 15 Medlar
16 Quince 17 Apricot 18 Peach 19 Apple 20 Pear

Vegetables

Vegetables generally require less preparation and care than other crops, and are the main component of kitchen gardens all over the world. Some fruit and salad vegetables are now grown commercially in greenhouses. Spain is particularly successful at this. By exercising control over the growing conditions, farmers can cultivate these crops out of season. Other vegetables are also raised as cash crops. In Europe and the United States, for example, fresh peas have become a rarity. Almost all the harvest now goes to canning and freezing.

The most useful vegetables are the pulses, also known as legumes. This is a group of plants that includes peas, beans, lentils and soya beans. As well as providing much more protein than cereals, pulses also put nutrients back into the soil. One of the ways agriculture could be improved is to make greater use of these plants.

Other future advances will probably involve making use of new crops. These include the Winged bean, of which all parts can be eaten, and the Yeheb bush which produces nutritious seeds in arid conditions.

Some common tropical vegetables

Some common temperate vegetables

1 Aubergine 2 Okra 3 Breadfruit 4 Avocado
5 Bamboo shoots 6 Endive 7 Jackfruit

1 Chives 2 Shallot 3 Onion 4 Garlic 5 Leek
6 Tomato 7 Globe artichoke 8 Spinach 9 Lettuce
10 Rhubarb 11 Asparagus 12 Florence fennel
13 Chicory 14 Celery

Sugar and oil

Sugar

Table sugar only became commonplace about 300 years ago, when sugar cane was introduced into America. Before this time, honey was the major source of sugar, and bees were widely kept. Sugar cane is native to South-east Asia, but it is now grown in most tropical and sub-tropical countries. Growth is limited only by the availability of water.

Sugar cane grows in tightly packed stands that reach an average of 6 m in height. After harvesting, which is generally done by hand, the sugary sap is squeezed out in a rolling mill. The sap is then boiled to make molasses and treacle. Further processing is required to produce the familiar white sugar crystals.

About 66 per cent of world sugar production now comes from cane, but in Europe, Russia and the United States, sugar beet is also very important. Sugar beet is the result of a breeding programme of the 1700s to improve natural beets, and the plant now has a 20 per cent sugar content.

In some tropical countries, brown sugars are manufactured from dates and other palms, but these are mainly for local consumption.

Oil

From the earliest times, certain plants have provided oil for cooking and lighting. Until the 1800s, vegetable oils were almost the only type available to industry. Today, a very wide range of crops are grown for their oil. In all cases the oil is squeezed out of the seeds or fruit, and the residue used as animal feed.

The commercial farming of oil-producing plants received a tremendous boost from the invention of margarine towards the end of the 1800s. During the last 20 years, vegetable oils have received a further boost, as increased health consciousness has led many people to consume less animal fat.

In tropical regions of Africa and Asia, especially in Indonesia and Malaysia, the oil palm is widely cultivated for cooking oil. The oil extracted from dried coconuts is also a valuable commodity, but it is mainly used industrially.

In Mediterranean climates, the olive is the main oil producer. Like the date palm, olive trees require little attention for most of the year, but harvesting is very hard work. Olive oil is not widely used outside the producing areas because there are cheaper alternatives.

▲ In Spain, and other
Mediterranean countries, olive trees
are grown on hillside terraces for
their valuable oil. Olive trees are
very slow-growing.

◀ Sugar cane is harvested (far left)
and processed (left) at a sugar
factory in Barbados. After crushing,
the cane can be processed into
paper, and the molasses residue
makes excellent animal feed.

Oilseed rape is grown for the oil
in its seeds. Its oil is edible but has a
very unpleasant smell. The crushed
seeds go to make cattle-cake. It has
become a popular crop in Britain.

Fibre and beverage crops

Fibre crops

A few crops are grown primarily for the useful fibres that can be extracted from various parts of the plant. Vegetable fibres have been used for centuries to make cloth, canvas, carpets and rope. For many applications, however, they have now been replaced by man-made fibres, such as rayon and nylon.

Fibres from seeds, stems and leaves are all used. Some fibre crops also provide oil and animal feed. The most important fibre crops are grown in tropical and subtropical regions, often on large plantations.

Cotton is the world's leading fibre crop, and is grown in more than 60 countries. The major producers are the United States, Russia, China, Korea, India, Brazil and Egypt. The fibres are taken from the top of the plant, where they surround and protect the seeds in the cotton boll, or seed pod.

Since 1800, the process of extracting the seeds from the fibres, known as ginning, has been mechanized. In other respects, cotton is still a very demanding crop, requiring constant attention. In the United States, Mexico and Russia, the cultivation of large fields is aided by machines. Elsewhere, cotton cultivation requires a large labour force.

Until very recently jute was second only to cotton in importance. Jute is a tall plant, 2-5 m high, that produces strong fibres in the stem. Today, jute fibre has mainly been replaced by plastic, but it is still important for twine, and in the upholstery and carpet industries.

Flax is grown both to produce the fine tough fibre of linen cloth, and for linseed oil. The plant grows well in cool, moist climates, and Europe and Russia are the major producers.

Several plants from the agave family are grown for the tough fibres in their leaves. The most important yield the fibres known as sisal, which is mainly grown in Central America, but is also cultivated in East Africa and Indonesia.

▶ A worker in Uganda picking the ripe red berries from a coffee bush. Each berry contains two seeds, or beans. Machines remove the pulpy flesh surrounding the beans, which are then dried. Next the hulls, or coverings, of the beans are removed, producing what is called green coffee. Roasting the green coffee beans allows them to develop their rich flavour.

Other fibre crops include hemp, which wa once grown for rope; kapok, collected from th seeds of a tropical tree; and ramie, a gras grown in China and used both for clothing an for making paper.

Coffee and tea

Coffee and tea have no food value, yet they ar two of the world's most valuable agricultura commodities. Several countries depend on on or other of these crops to maintain the national economies.

Our 300-year-old appetite for morning coffe is now met largely from South America, an Brazil alone supplies about half the world consumption. Coffee is also widely cultivated i East Africa. After harvest, the bright red coffe berries must be dried and skinned in order t reveal the coffee "beans". Most of the beans ar now processed and sold as instant coffee.

Tea originated in South-east Asia, where has been drunk for thousands of years. It wa introduced into Europe in the 1700s. Toda India and Sri Lanka together produce about 8 per cent of world production. Large quantitie are also exported from Indonesia and Ea Africa. Growing on the bush, tea leaves are 5-cm long. These are crushed and dried befo being packed for shipping.

▲ In Peru, South America, a cotton-picker sorts through a mass of freshly picked cotton, removing twigs and leaves. The fluffy "snowballs" of cotton still contain the seeds, which are later removed during ginning. Yarn is spun from long cotton fibres, called lint. Short fibres, called linters, are used to make cotton wool, and are a useful raw material for making rayon.

◀ Picking tea on a large plantation in Tanzania, East Africa. The tea plant is a small shrub, which is kept well pruned so that it keeps producing fresh young leaves. The picked leaves are allowed to wilt for a day or two before being crushed. Then they ferment in the air, which makes them turn black. The process is stopped by heating them in an oven.

Cattle farming

Spot facts

- *In the United States, over 90 per cent of all cereal production is used as animal feed.*

- *A single cow can provide up to 4,000 kg of milk per year.*

- *Rabbits will turn the same amount of vegetation into meat as cattle, but will do it four times more quickly.*

- *Bulls from Chianini in Italy average 850 kg in weight. Those of Ovambo in Namibia, West Africa, average only 225 kg.*

- *The world's largest cattle ranch, in southern Australia, covers 30,113 sq km, which is 23 per cent of the area of England.*

▶ Zebu cattle on a ranch in Brazil. Originating from India, zebu are well suited to subtropical climates. The distinctive hump is thought to have been selected for by centuries of use as draught animals. Yoking the cattle is made easier by the hump. In India, the cow is considered a sacred animal and is not exploited for meat.

Cattle are the world's most important agricultural animals, providing high-quality meat, milk and muscle power. Beefsteak is widely considered to be a luxury food, and in some less developed countries, cows are still an important symbol of wealth and status. Cattle have least agricultural importance in the rice-growing areas, and in India they are used only for milk and as draught animals. Both beef and dairy farming became global industries when canning and refrigeration were invented during the 1800s. By 1900, beef and butter were being traded around the world. One of the major uses of beef in the West is for hamburger, for which top-quality meat is not needed.

Raising cattle

There are three basic kinds of cattle: longhorn, shorthorn and humped zebu. All share a common ancestor, the auroch, which is now extinct. Of the surviving breeds, longhorns bear the closest resemblance to primitive cattle.

Today, commercial cattle farming is either intensive, as in northern Europe, or extensive, as on the cattle ranches of the American West. In East and West Africa, sizeable numbers of cattle are kept by nomadic herders.

Cattle ranching requires large areas of land with year-round natural grazing, and was confined to temperate grasslands. Ranching is now established in South and Central America on land cleared from rain forest.

Methods of cattle rearing

A rancher needs to be extremely mobile, hence the partnership between people and horses. The 'cowboy" has existed in various parts of the world for over a thousand years, first in North Africa, then later in Spain, America and Australasia. However, the modern cowboy is as likely to ride a motorcycle as a horse.

Where land is at a premium, cattle are raised intensively. In countries such as Britain, Holland, Denmark and Germany, cattle are kept indoors during the winter and fed on stored fodder such as hay. Modern intensive production requires that the animals are kept indoors for most, if not all, of the year. They are fed a combination of cattle-cake, grain, root crops and cabbages. In the United States many cattle are raised intensively, especially near big cities.

The ratio of cattle to land varies greatly. In Australia, where the grazing is quite poor, the average density is only four animals per square kilometre of land. In the United States, where both intensive and extensive methods are used, the average density is the same as that for the whole world, at 25 animals per sq km. In Holland, cattle are raised at an average density of 180 animals per sq km.

► Young beef cattle being fattened up for sale on a farm in California, in the United States. When cattle are raised intensively for meat, the animals themselves require very little room or human labour. However, in terms of the land required to grow their feed, the cattle indirectly occupy a much larger area. There is also the problem of disposing of the animals' waste.

Dairy cattle

Cattle provide us with more than eight times as much food in the form of milk than in the form of meat. Nearly all commercial milk production comes from intensive farming. Because cows require milking twice daily, dairy cattle cannot be spread over large areas. Only in New Zealand is the grazing of sufficient quality to allow extensive dairy farming.

Cattle are raised for milk in most parts of the world, often on the basis of one cow per village. The main areas of commercial dairy production are in Europe and North America. The traditional dairy herd was quite small because milking by hand is a very slow and tiring process. Ten cows were all that a single person could handle.

The invention of the milking machine in the 1940s allowed a single farm worker to milk up to 80 cattle. The modern milking parlour, together with the introduction of year-round shelter and improved animal feed, has transformed dairy farming into an extremely productive industry.

Most breeds of cattle can be raised for both beef and milk, but shorthorn dairy breeds give much higher milk yields.

▼ The best dairy breeds provide excellent milk but are much too skinny for commercial meat production. The Brown Swiss has been used to improve dairy herds in the United States. The Jersey is a small breed that gives very creamy milk and makes an ideal "family" cow. Finn and Normandy breeds are popular in the USA.

Milk is a high quality food, rich in protein fats and essential minerals. Through selective breeding, agricultural scientists have recently improved both the quantity and quality of milk produced. Compared with her counterpart of 40 years ago, the modern dairy cow produces over twice as much milk, with a higher fat content.

The best quality milk is generally used for butter and cheese making. These milk products are a much more concentrated form of food, and are more easily stored than fresh milk which goes sour very quickly.

Britain makes more use of fresh milk than any other country. Nearly 70 per cent of total milk production is consumed in this way. By contrast, less than 10 per cent of New Zealand's milk is drunk as a liquid; most is turned into butter for export. In the developed countries as a whole, about 15 per cent of production is made into cheese.

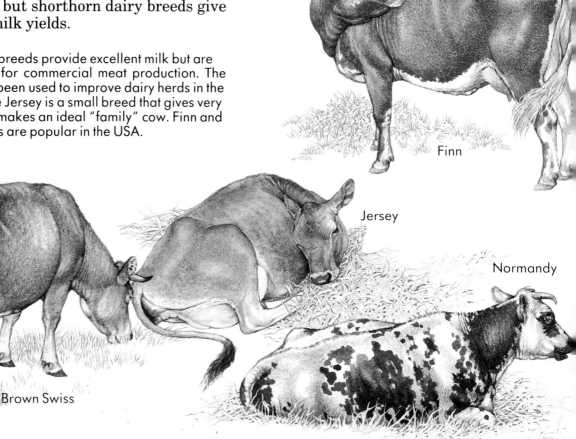

Finn

Jersey

Normandy

Brown Swiss

▶ A modern circular milking parlour in Scotland. The cattle are milked by sophisticated equipment while they stand on a rotating platform. The farmer can adjust the time of rotation to the average length of milking. This arrangement is expensive but very efficient because it permits a steady flow of cattle through the parlour. Only one operator is needed.

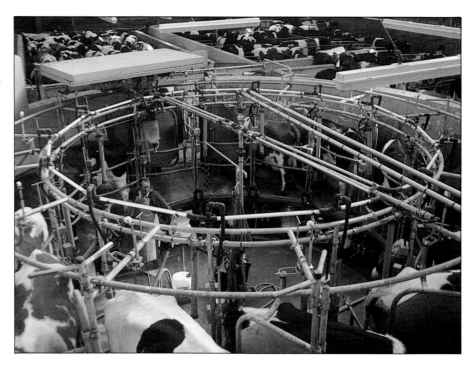

Milk produced by each species
(in million tonnes)

Goat 7.7
Sheep 8.1
Buffalo 28
Cow 438

Milk production

Milk is collected into graduated recording jars and kept cool in a farm vat. It then goes by hose into a 9000-litre milk tanker. From there it may be loaded into a 20,000-litre tanker and taken to the processing dairy. Milk is pasteurized to destroy bacteria, and is then processed into the many milk products now available in the shops and by doorstep delivery.

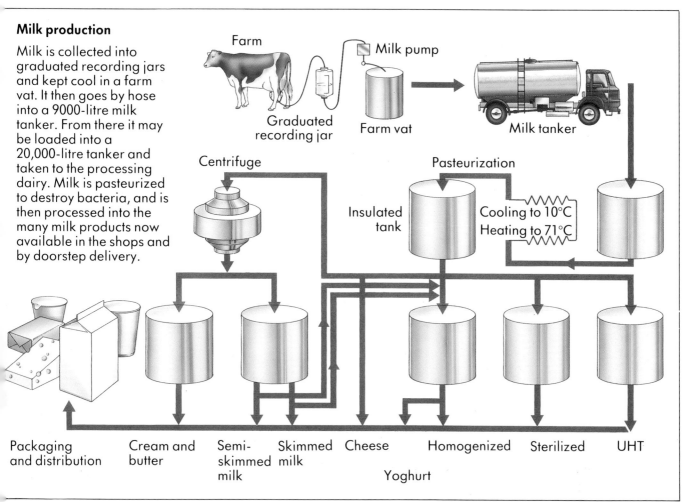

Farm
Milk pump
Graduated recording jar
Farm vat
Milk tanker

Centrifuge
Pasteurization
Insulated tank
Cooling to 10°C
Heating to 71°C

Packaging and distribution
Cream and butter
Semi-skimmed milk
Skimmed milk
Cheese
Homogenized
Sterilized
UHT
Yoghurt

Beef cattle

Beef and veal are the most expensive forms of meat to produce, yet they are so popular that they account for one-third of world meat consumption. The taste for beef is concentrated in the developed world. In most other regions cattle are too valuable to be killed for food.

Cattle are not particularly efficient meat producers, and only about 60 per cent of the animal is edible. Beef cattle require high-quality food in order to put on weight, and in Europe and North America they are raised almost entirely on concentrated feeds.

In the developed world, only beef from cattle that have been fed on grain or cattle-cake is considered fit for the table. Meat from cattle that have been ranched on grassland is generally of lower quality and is only used in hamburgers and processed meat products.

European breeds produce the best-quality meat, but cannot tolerate warm climates or poor grazing. The huge herds that are now ranched in South and Central America are largely composed of longhorn and zebu cattle that have been improved by crossbreeding with European stock.

In parts of the United States and Europe, beef cattle are now raised very intensively, and some animals, especially those destined for veal, spend their whole life indoors.

Auroch

Hereford

Simmental

Blue Belgian

▶ Modern beef cattle are not as sturdily built as the ancient auroch, but some of them approach it. The Hereford is the main British meat-producer, and has also been crossbred with American stock. The Simmental and Blue Belgian breeds both achieve great size and reflect attempts to produce the most meat possible.

▶ (opposite) Black Welsh cattle are smaller than most beef cattle. Being strong and sturdy, they are well suited to their native upland terrain.

Other livestock

▶ The nomadic peoples of southern Central Asia and Mongolia obtain meat, milk and clothing from their flocks of goats. During a year, the flocks may be herded hundreds of kilometres between summer and winter pasture. In mountainous regions, the flocks are brought down to the valleys for the winter, to avoid the worst weather.

Very few other animal species are important in world agriculture. Pigs are the most efficient meat-producers, and it was the boast of some American meat-packers that no part of a pig was wasted.

In general, sheep and goats are the most versatile livestock, providing meat, milk and wool. Goats are also extremely hardy and are found in mountain ranges and on the fringes of deserts.

Poultry have long been a source of meat and eggs. During the 1900s, poultry farming has become a huge industry in the developed world. In terms of increasing productivity, chicken is the fastest growing foodstuff in the world.

In many parts of the developing world, horses, mules, donkeys, camels, and, in South America, llamas, all have a limited role in agriculture.

Pigs

Wild pigs are found in many parts of the world and their domesticated cousins, known as hogs in America, are the world's major meat-producers. Each year, a world population of about 750 million pigs produce 55 million tonnes of meat. With the best breeds, up to 80 per cent of the animal can be turned into food.

Pigs are natural scavengers; they will eat almost anything and need little attention. In Asia, many families buy one or two piglets each year to fatten up on kitchen scraps. In Europe and North America, which have about half the world's pigs, most of the animals are reared intensively and indoors. Where possible, many farmers fatten their pigs on seasonal fruits such as acorns.

Pigs are well suited to mass-production methods, and modern farmers aim to get at least 20 piglets per breeding female per year. When fed on grains and protein supplements, each piglet will grow about 65 kg of usable meat in just six months.

Pig meat deteriorates rapidly in hot climates, and some cultures consider it unclean. In cooler climates, pork is ideally suited to traditional forms of preservation, such as smoking and salting, to produce ham and bacon. Large, solidly-built breeds are often referred to as "bacon" pigs. Other traditional breeds, often known as the "coloured" breeds, were raised to produce meat with a lot of fat. Chinese pigs also produce fatty meat. In the developed world, public taste now favours lean meat from leaner pigs such as the Landrace.

Meat from different animals, 1981 (million tonnes per year)

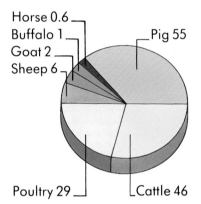

Horse 0.6
Buffalo 1
Goat 2
Sheep 6
Pig 55
Poultry 29
Cattle 46

◀ The pig is the world's number one meat producer, accounting for about 40 per cent of the total annual production of 140 million tonnes. Cattle are the next most important (33 per cent), followed by poultry (21 per cent). All other types of meat together make up the rest.

1 Landrace

2 Duroc

3 Berkshire

4 Andalusian

5 Craon

▲ (1) Landrace pig from Denmark, a lean pig that has recently increased in popularity. Traditional European breeds such as (2) Duroc, (3) Berkshire, (4) Andalusian, and (5) Craon were bred to be hugely fat, and are now less popular.

Sheep

The sheep is a multi-purpose animal that provides meat, milk and wool. Sheep are raised throughout the world, and have the greatest commercial importance in Australasia, Europe and Argentina. In New Zealand, for example, sheep outnumber people by a ratio of 20 to 1. In Britain, 90 per cent of sheep are raised for meat, nearly all of it in the form of lamb. In other countries, milk for yogurt and cheese-making and wool are the major products.

Sheep are hardy animals that perform well on a diet of grass, and may be kept outdoors all year round. Their heavy woollen fleece is naturally waterproof, and provides excellent insulation. Apart from shearing and the lambing season, sheep do not require much attention. A flock of 500 may be tended by a single shepherd and a dog. Having a thick fleece makes sheep very attractive to ticks and other parasites. Most developed countries require their sheep to be dipped regularly in pesticides. Sheep that have been dipped are marked with a brightly coloured dye.

There are hundreds of different varieties o sheep that have been bred to take advantage o different types of grazing. In general, upland sheep are closer to the domestic sheep' primitive ancestors. Upland breeds provide les wool and meat per animal, but will graze on windswept hillsides that are too cold and exposed for their more refined downland cousins. Downland sheep require more attention, but are well suited to the production o quick-growing lambs for meat.

There are many local breeds of long-woolled sheep, but the world's dominant wool produce is the Merino. The breed originated in Spain and was taken by colonists to South America South Africa and Australasia. By 1900, Aust ralia had over 100 million sheep, virtually all o them merinos. Sheep are second only to cattl as a source of animal products. World outpu totals 115 million tonnes of sheep products, o which only 6 million tonnes is meat. Th average British sheep produces about 5 kg o wool per year, from one shearing.

▶ Long-woolled sheep on a French mountainside.

▼ (1) Sudanese sheep. The Merino has been crossbred with many local sheep, (2) is the Arles variety. (3) The West African dwarf breed inhabits the humid forest zone, and the Soay (4) lives in the Outer Hebrides.

1 Sudanese

2 Arles

3 West African dwarf

4 Soay

Sheep-shearing is hard and very skilled work. Fleeces weigh an average of 12 kg each, and the huge size of some flocks (up to 50,000 animals) means that each shearer has to deal with 40-50 animals per hour. Scientists are now working on producing a self-shearing sheep that would shed its fleece after a single injection of hormones.

Goats

Goats are closely related to sheep, and are agile and hardy animals. The more primitive breeds of both species closely resemble each other, and in the warmer parts of the world they are often raised in mixed flocks. Like sheep, goats provide highly nutritious milk that is often made into cheese. Goats are also important for wool and leather. In total the world uses about 450 million tonnes of goat products each year (about one third the quantity of cattle or sheep products), of which only 2 million tonnes are meat.

In general, goats can thrive in a more arid climate, and on lower-quality grazing, than sheep. In this respect they are extremely useful because they can take advantage of land that would otherwise be useless for agriculture. But goats can also be very destructive.

A combination of agility and wide ranging appetite means that without careful management, goats will devour all the available vegetation, which in turn encourages soil erosion. They are usually kept tethered.

Some breeds of goat tolerate the cold better than other domestic animals, and mountain peoples often rely entirely on goats for both food and clothing. Some Asian breeds are especially noted for their high quality wool. Cashmere is obtained from breeds of goat originally from northern India. Each goat provides only half a kilogram of fibres each year. Mohair is the hair of the Angora goat, originally from central Turkey. Both breeds are now farmed.

In Europe and North America, goats are kept for milk production and are often raised intensively on concentrated feeds. The Swiss goat is the traditional milking breed, although during this century it has been challenged by a hybrid strain, the Nubian goat.

A cross between the small English goat and a larger Egyptian breed, the Anglo-Nubian, was first introduced around 1900. By 1950, half of all goats raised in the United States were Anglo-Nubians. In Europe, the Swiss goat remains dominant, giving more milk, but of a lower quality by comparison.

▶ The agility of goats is legendary. Goats appearing to climb trees are a familiar sight in warm countries.

▼ The commercial production of goat's milk can be mechanized in the same way as that of cow's milk.

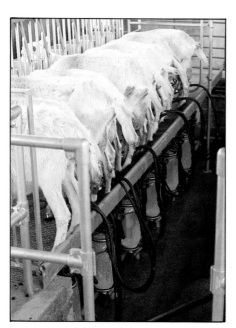

Nordic

Anglo-Nubian

Saanen

French Alpine

British Alpine

► Breeds raised for milk production include: Nordic, from northern Scandinavia; Anglo-Nubian, a common sight in North America; French Alpine, which is found in most parts of France; Saanen, originally from Switzerland and now farmed all over the world; British Alpine, a cross between English and Swiss breeds; Mamber, from the Syrian mountains, a typical Middle-Eastern breed; and Murcia-Granada, which is found only in southern Spain.

Mamber

Murcia-Granada

Poultry

Chickens dominate the world's poultry population; 97 per cent of all farmed birds are chickens. In the Western world, chicken is rapidly becoming the most popular form of meat. In 1950 Britain consumed less than 5 million chickens. Forty years later, the total has increased to 500 million, and is predicted to reach 1,000 million by the year 2000. Each year the world consumes about 30 million tonnes of poultry meat, and an equivalent weight of eggs.

In the developed world, laying hens and those destined for meat (broilers) are usually raised intensively. The birds are kept in large buildings containing thousands of individual cages arranged in stacks (batteries). They are fed carefully controlled amounts of food by a conveyor-belt system. By the use of artificial lighting, the hens' natural egg-laying cycle can be altered so that they provide a constant supply of eggs.

In Britain and the United States, turkeys are raised intensively for seasonal consumption, but are slowly becoming a year-round food. Geese, however, are declining in popularity. In Asia, ducks are often raised on flooded rice paddies. Duck meat and eggs are used.

▲ Turkeys in Wisconsin, USA, being fattened on expensive grain for the Thanksgiving table. The mass-production of poultry demands very large numbers of birds, and turkeys are not suitable for confinement in batteries. Open-air raising brings risks of injuries from fighting.

◄ Ducks on the Pacific island of Bali being driven to market. Ducks are one of the most widespread bird families, and throughout the world local species have been domesticated for egg and meat production. They require little attention, but do need a pond.

► Chicks hatching in an incubator. Efficient poultry production requires control over every stage of the process. An incubator permits the eggs to hatch under the most ideal artificial conditions, thus ensuring that the greatest possible number of chicks survive.

◄ Toulouse geese are among the largest domesticated birds.

▼ Birds outnumber mammals two to one. Each domesticated mammal yields about 30 kg of meat per year. Each bird yields only 9 kg of protein, half of it in the form of eggs.

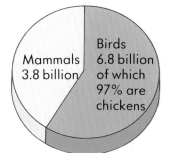

Mammals 3.8 billion

Birds 6.8 billion of which 97% are chickens

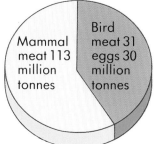

Mammal meat 113 million tonnes

Bird meat 31 eggs 30 million tonnes

Fishing

• Between 1964 and 1975, the catch of North Atlantic haddock fell by more than 90 per cent as a result of fishing methods that were too efficient.

• A modern factory ship can process over 1,000 tonnes of fish per day, and may remain at sea for up to nine months.

• One third of all the fish caught off the West African coast are used as fertilizer and animal feed in the developed countries of the world.

• A Chinese fish farm can produces 4.5 tonnes of animal protein per hectare per year, 10 times more than if the same area of land were used for mammal livestock.

▶ Fish caught in the trawl net being hauled aboard a side trawler. Harvesting the sea is now as mechanized as harvesting the land. But in agricultural terms, our approach to the oceans is still primitive. The crop of wild fishes is steadily declining in the face of over-fishing. More and more of the fish we eat comes from fish farms.

Fishing represents the only significant source of human food that still relies on natural food chains. Our attitude to the sea is still largely that of the hunter-gatherer.

On average, the world gets about 6 per cent of its protein from fish and shellfish, but in some countries, including Japan, South-east Asia, Portugal, Norway, Russia, and some West African countries, the proportion is considerably greater. About a third of the total marine catch is not used as human food, but is converted into animal feed and fertilizer.

Modern fishing methods are extremely efficient, and over-fishing is now the industry's greatest problem. Part of the response has been a revival of fish farming.

Fishing grounds

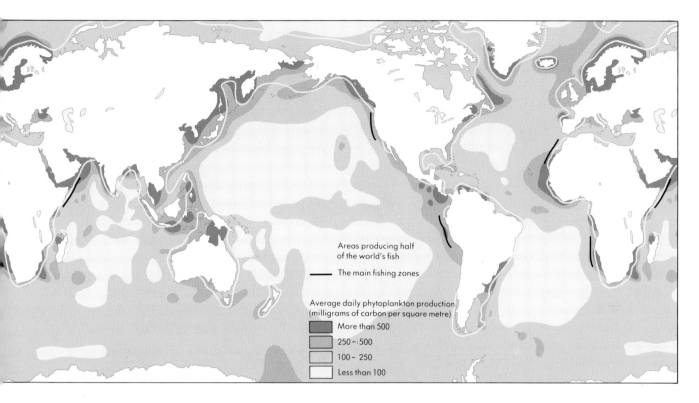

Areas producing half
of the world's fish

—— The main fishing zones

Average daily phytoplankton production
(milligrams of carbon per square metre)

More than 500

250 – 500

100 – 250

Less than 100

▲ Microscopic phytoplankton produce food from sunlight using photosynthesis. Marine scientists can calculate the food content of seawater by measuring the quantity of carbon in the phytoplankton.

▼ Tiny shrimp-like krill occur in great numbers in the oceans around Antarctica. Russia catches about 1 million tonnes per year. The total sustainable harvest may be as high as 50 million tonnes.

Most of the fishes we eat stand fairly high in the ocean food chains. They occur in the greatest numbers in coastal waters, where plant life is most abundant. It is warmer here, and there is a run-off of nutrients from the land.

Offshore fish populations are concentrated in fairly shallow waters (less than 150 m) above the continental shelf where concentrations of phytoplankton are greatest. The richest fishing grounds are often associated with the upwelling of deep ocean currents that carry nutrient-rich waters from mid-ocean.

One of the world's most productive fishing grounds, off the coast of Peru, was fed by the El Niño current that brought plankton from the Central Pacific. During the early 1970s, the Peruvian fishing industry landed over 13 million tonnes of fish per year. A sudden change of direction in the current deprived the area of most of its food supply. Within ten years, catches had dropped to less than 2 million tonnes of fish per year.

Catching the fish

The richest fishing grounds within easy reach are coastal shallows, lagoons and river estuaries. Some rivers are used by migratory fish, such as the salmon, and provide an additional seasonal harvest. Many shellfish, for example shrimps, are also plentiful during the spawning season and can be collected with little effort.

Shallow waters, up to 15 m, can be fished from small boats powered by oars or paddles. A wide variety of methods are still used for subsistence fishing. Nets, baited hooks, spears, even trained seabirds such as cormorants, are employed to catch fish. But only large nets, handled by many fishermen working together, are capable of landing a worthwhile surplus which can be sold.

Most commercial fishing takes place offshore, where pelagic and demersal fish account for over 75 per cent of the total catch. Pelagic fish include herring, sardines, and tuna; they occur in free-swimming shoals at depths up to 100 m. Demersal fish, such as cod and plaice, are bottom-dwellers found on the continental shelf and offshore banks.

Each year we haul about 75 million tonnes o marine life out of the sea to use as food. Some 6(per cent of the total catch is landed by just 1(countries, with Japan and Russia each accoun ting for more than 10 per cent of the harvest.

World fish production has remained stati during the last 20 years despite the increase use of modern fishing fleets and factory ship that can process the catch in mid-ocean. Popula tions of the traditional food fishes have bee falling dramatically because of over-fishing Only by switching to unfamiliar species, man of which are used only for animal feed an fertilizer, has the volume of the catch bee maintained.

Pollution has also affected the quality of th catch. Some important species have been foun to have acquired dangerous levels of poisonou pollutants through bio-concentration.

▶ Shore fishermen in Malaysia. One major disadvantage of shore fishing with a net is that large fis are fairly rare. The average size of the fish caught is usually very small (less than 15 cm).

▼ These are important species of commercial marine fish. These species swim well clear of the seabed and are found in wide ranges of seas. The limits of their territory are set by the temperature of the water.

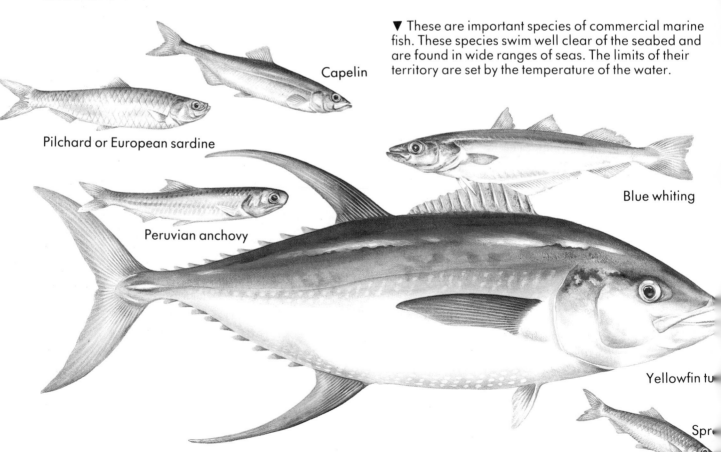

Capelin

Pilchard or European sardine

Peruvian anchovy

Blue whiting

Yellowfin tu

Spr

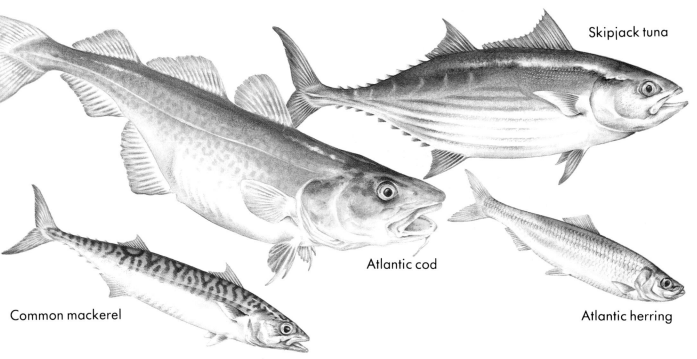

Skipjack tuna

Common mackerel

Atlantic cod

Atlantic herring

Netting methods

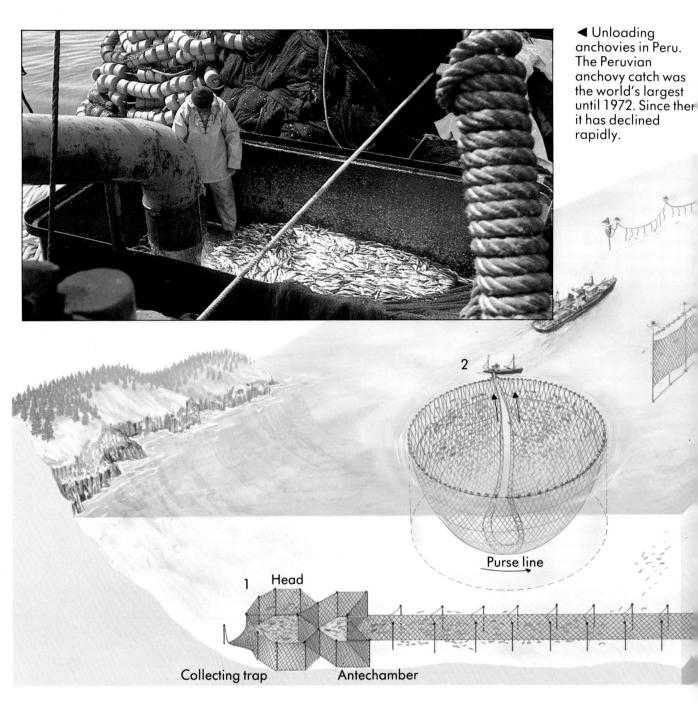

◄ Unloading anchovies in Peru. The Peruvian anchovy catch was the world's largest until 1972. Since then it has declined rapidly.

2

Purse line

1 Head

Collecting trap   Antechamber

Top twenty fish

A fairly small number of species dominate the sea's harvest. The Alaskan pollack and the Japanese pilchard together make up over 10 per cent of today's catch. Some commercially important species are completely unknown to most people. For example, the Gulf menhaden which makes up 40 per cent of the United States' catch, is not considered fit for human consumption. The entire catch of menhaden is processed into animal feed.

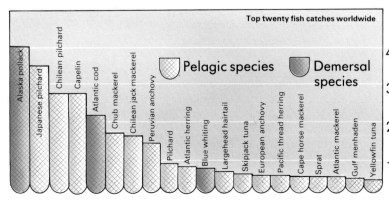

Top twenty fish catches worldwide

Alaska pollack
Japanese pilchard
Chilean pilchard
Capelin
Atlantic cod
Chub mackerel
Chilean jack mackerel
Peruvian anchovy
Pilchard
Atlantic herring
Blue whiting
Largehead hairtail
Skipjack tuna
European anchovy
Pacific thread herring
Cape horse mackerel
Sprat
Atlantic mackerel
Gulf menhaden
Yellowfin tuna

Pelagic species   Demersal species

4

3

2

1

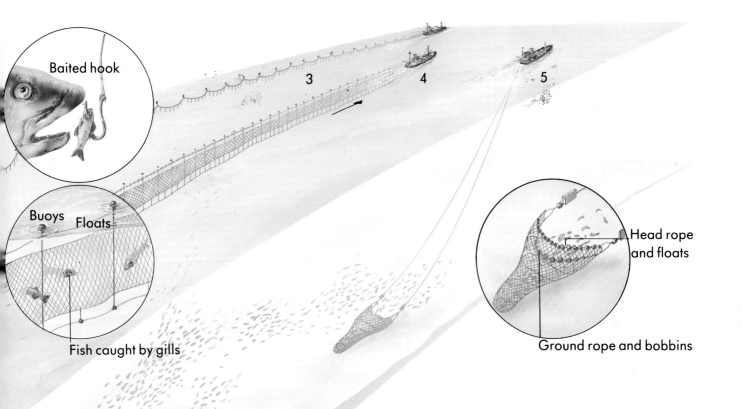

Baited hook

Buoys Floats

Fish caught by gills

3

4

5

Head rope and floats

Ground rope and bobbins

The basic technology of fishing is very simple, and the vast majority of fish are caught in nets from boats. Other methods, such as netting fish from the beach or trapping shellfish in baskets, have only local importance. In some estuaries and straits, complex arrangements of fixed nets are employed to lead fish into a fixed collecting trap, which is regularly emptied.

Trawling and seining are the two most important sea-fishing techniques. Today, the fish are actively pursued, often by means of echo-location equipment. A trawl net may be drawn by one or two ships. If a single ship is used, otter boards are fitted to keep the mouth of the net open. Both pelagic and demersal fish are caught by trawling.

A seine is a single net that is drawn around a shoal of pelagic fish like a purse. Some seine nets are over 1,000 m long. Seining was developed from drifting, in which a net is merely trailed behind a slowly moving boat. Floats and weights are fitted to the net to keep it up-right in the water.

Some fish, especially cod, are caught on a commercial scale on hooks. Trailed from a fishing ship, a single long-line might hold 5,000 individually baited hooks.

▲ (1) Fixed net (2) Seining (3) Long-lining (4) Drifting (5) Trawling. Over-fishing is not only about numbers of fish. In all forms of net fishing, the size of the individuals caught can be controlled by varying the size of the mesh in the net. Taking too many small (young) fish causes the population as a whole to decline.

195

Farming the waters

▶ An inland fish farm in Arizona, USA. Linking the ponds allows a flow of clean water.

▼ A French oyster farm at low tide. When the tide comes in, the shellfish can feed on nutrients in the water.

▼ These are some of the more widely cultivated vertebrate fish. Catfishes are bottom-feeding scavengers, and carps are herbivorous. Herbivorous species can be fed more cheaply than carnivores and tend to be more highly productive.

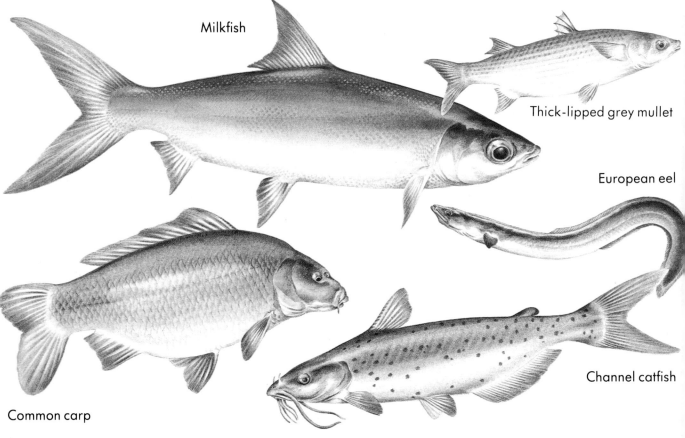

Milkfish

Thick-lipped grey mullet

European eel

Channel catfish

Common carp

Fish farming, the breeding and raising fish like farm animals, is a continuing tradition in much of Asia. The technique has achieved its highest sophistication in China, where a freshwater fish pond can produce 10 times as much protein as the same amount of farmland.

A Chinese fish pond is a complete ecosystem in which all the species are edible. The upper and middle levels of the pond are occupied by various species of carp. Their diet of water plants and algae is supplemented by waste vegetation. At the bottom of the pond, dace feed on the detritus and keep the water clean. Fish are also raised in flooded paddy fields, and are shifted from one field to another when the time comes for rice planting and harvest.

Throughout the Pacific, and especially in Indonesia, thousands of hectares of coastal mangroves are farmed for marine species. Milkfish and grouper are raised in large ponds, and shrimps are harvested regularly like crops.

In other parts of the world, the traditional fish ponds fell into disuse centuries ago. Until very recently, fish farming has been confined to some shellfish such as oysters and lobsters.

During the last 20 years there has been a tremendous revival in fish farming, especially in the United States and Europe. One quarter of British fish consumption now consists of trout and salmon produced on farms. Trout are raised in ponds or enclosed stretches of river. Salmon are farmed in estuaries and sea lochs in Scotland. The fish are contained within huge floating cages as large as a football field. A single farm may have over 500,000 live fish at various stages of growth. All the food for both salmon and trout has to be provided by the farmer.

Fish kept in enclosed conditions are extremely vulnerable to attack from disease and parasites. Western fish farmers use large quantities of powerful pesticides to protect their fish. There is mounting concern about such substances being released directly into marine ecosystems.

▼ Netting salmon at a saltwater fish farm in Washington State, USA. After harvest, the first task is to remove eggs which are used for raising the next generation. Young salmon are initially raised in fresh water, as in nature.

Nature in balance

● As many as 30 million different species of plants and animals may live on Earth. Only 1.7 million have yet been named, and only a few thousand have been studied closely.

● A single 30-cm-long fish represents the food energy of perhaps 100 million plankton.

● One hectare of rain forest may contain up to 200 different tree species. In the coniferous forests, one tree species may be the only tree to cover hundreds of square kilometres.

● A lion needs to kill and eat about 50 zebra (or similar-sized animals) each year, in order to survive.

▶ City schoolchildren pond-dipping in Essex, England. Environmental education is becoming more common in many countries. Concern is growing about the many threats to wildlife and the balance of nature posed by the growth of industry and human populations. Young people are especially keen to play their part in restoring the natural balance.

No living organism exists in complete isolation from others. Plants and animals that live in a particular place share the same air, the same rocks and the same neighbours. They live together, and they live in harmony. Some feed on others, but the larger picture is one of overall balance.

Scientific studies have shown that this balance is achieved through an extremely complex network of relationships between different species. One of the most important relationships is that between plants and animals. Without plants, animal life would be impossible.

Ecology

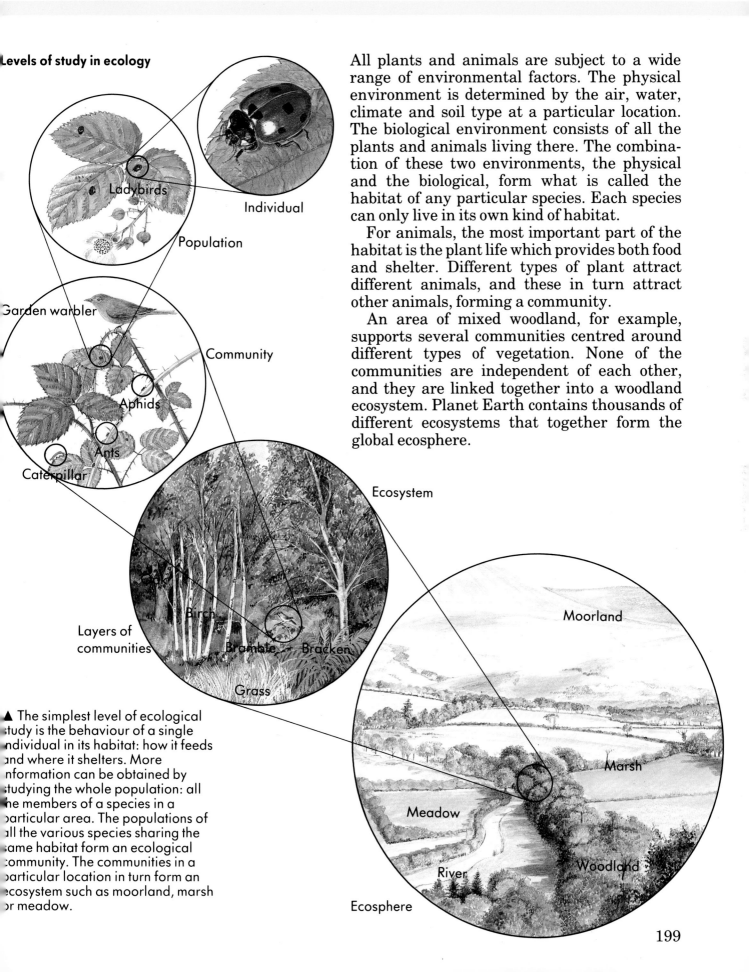

Levels of study in ecology

Ladybirds

Individual

Population

Garden warbler

Community

Aphids

Ants

Caterpillar

Ecosystem

Layers of communities

Birch

Bramble

Bracken

Grass

Moorland

Marsh

Meadow

River

Woodland

Ecosphere

All plants and animals are subject to a wide range of environmental factors. The physical environment is determined by the air, water, climate and soil type at a particular location. The biological environment consists of all the plants and animals living there. The combination of these two environments, the physical and the biological, form what is called the habitat of any particular species. Each species can only live in its own kind of habitat.

For animals, the most important part of the habitat is the plant life which provides both food and shelter. Different types of plant attract different animals, and these in turn attract other animals, forming a community.

An area of mixed woodland, for example, supports several communities centred around different types of vegetation. None of the communities are independent of each other, and they are linked together into a woodland ecosystem. Planet Earth contains thousands of different ecosystems that together form the global ecosphere.

▲ The simplest level of ecological study is the behaviour of a single individual in its habitat: how it feeds and where it shelters. More information can be obtained by studying the whole population: all the members of a species in a particular area. The populations of all the various species sharing the same habitat form an ecological community. The communities in a particular location in turn form an ecosystem such as moorland, marsh or meadow.

199

Themes in ecology

Ecology is the study of how species react to other species. By studying the relationships and interactions between species in a community, ecologists can learn how a balance is maintained in nature. The feeding behaviour of animals is one of the most important interactions. Plants can meet their own needs from sunlight, carbon dioxide, water and nutrients in the soil. Animals, however, have actively to search for food, since they cannot make their own.

An area of woodland may support a population of millions of flying insects. These offer a rich source of food for airborne predators able to catch them on the wing. This role is filled by certain birds and bats. A position within an ecosystem such as this is called a niche.

Animals may feed on the caterpillars of the same insects, but occupy a different niche. Similarly, the birds and bats occupy different niches because the birds hunt by day and the bats by night. This separation of niches avoids unnecessary competition for food. In general, each potential niche tends to be occupied by only one or two different species. Any more would make an imbalance.

Feeding behaviour is central to defining a niche, but other interactions are also very important to the ecosystem. Many animals are essential to plant reproduction. Insects fertilize flowers while feeding, and birds and mammals are used by plants to carry seeds great distances. Some interactions are extremely specific, and occur only between two species; others involve many species.

▲ The African Bushveld elephant shrew specializes in eating termites, an abundant source of food. The behaviour of any species is largely determined by its habitat.

◄ The giraffe occupies a very successful niche on the African plains. Its long neck enables it to feed on vegetation that is above other animals' heads.

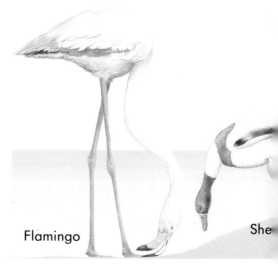

Flamingo   She

Succession

One year

Five years

25 years

When new land is created, or old land is cleared, a new community is born. The first living colonists will be plants suited to the open conditions. Plants are vital in forming and shaping a habitat. Their roots hold the soil together, and a mass of vegetation can moderate the effects of harsh climates. As the first community develops, it creates the shelter and shade necessary to nurture the next group of less hardy plants.

This progressive occupation of land is known as succession. Different types of vegetation follow distinctive patterns of succession. It may take hundreds of years for plants to turn these sand dunes into firm ground. After one year, grasses have already taken root. After five years, shrubs and small trees are established. After 25 years, the process is well under way, though far from complete.

Ecosystems do not suddenly come into existence; they are the result of many centuries of evolution. A community, on the other hand, can form during a much shorter period of time. Usually, the community will go through a series of stages before reaching its final form, which is known as a climax community. Whatever the ecosytem – prairie, tundra, woodland or jungle – a climax community will regenerate itself indefinitely under natural conditions. If a small part is damaged, it soon builds up again.

The first inhabitants of a community, which are known as pioneer species, arrive as wind-borne seeds and flying insects. On coasts and islands they may also be washed up by the sea.

At first there are relatively few interactions. As the community develops, other species may displace the pioneers, only to be displaced themselves as succession continues. Climax communities usually contain a wide variety of plant or animal species, linked by a very complex network of interactions.

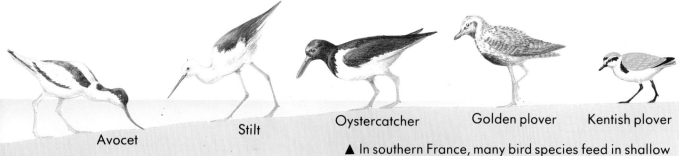

Avocet

Stilt

Oystercatcher

Golden plover

Kentish plover

▲ In southern France, many bird species feed in shallow lagoons. The length of beak, neck and legs determine the depth of water each can reach, and therefore the feeding niche of each species.

Food webs

The source of all food and living energy is the Sun. By studying feeding habits, ecologists can trace the flow of energy through an ecosystem. The process of photosynthesis enables plants to manufacture their own food. For this reason, plants are called primary producers.

Animals that feed directly on plants (herbivores) are known as primary consumers. Animals that obtain their food from other animals (carnivores) are called secondary consumers. These different levels of feeding can be linked together into a food chain which always begins with a form of vegetation. The stages in the chain are often described as belonging to different trophic levels.

Food chains link together individual plant and animal species. The various food chains within an ecosystem can be combined to create a much more complex food web. At the summit of the food web are the top predators, whose food energy may have passed through as many as five different organisms. The final connections in the food web are the decomposers. These are the bacteria and fungi that break down dead organic matter into its component parts and return the nutrients to the environment.

Food web in a temperate lake

- First trophic level (primary producers)
- Second trophic level (herbivores)
- Third trophic level
- Fourth trophic level
- Fifth trophic level
- Sixth trophic level

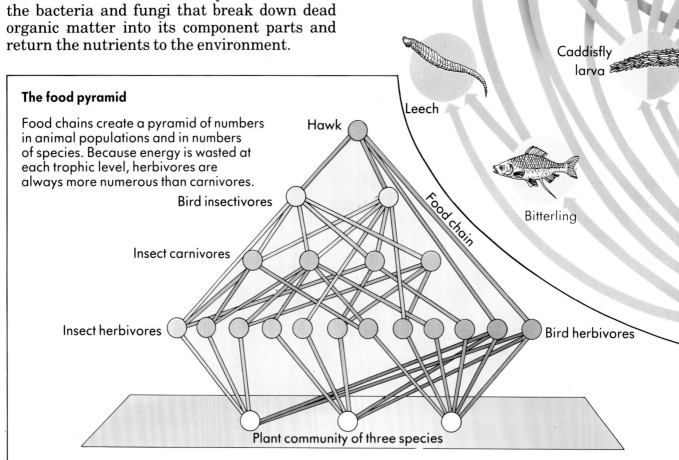

Duck

Frog/tadpole

Leech

Caddisfly larva

Bitterling

Food chain

The food pyramid

Food chains create a pyramid of numbers in animal populations and in numbers of species. Because energy is wasted at each trophic level, herbivores are always more numerous than carnivores.

Hawk

Bird insectivores

Insect carnivores

Insect herbivores

Bird herbivores

Plant community of three species

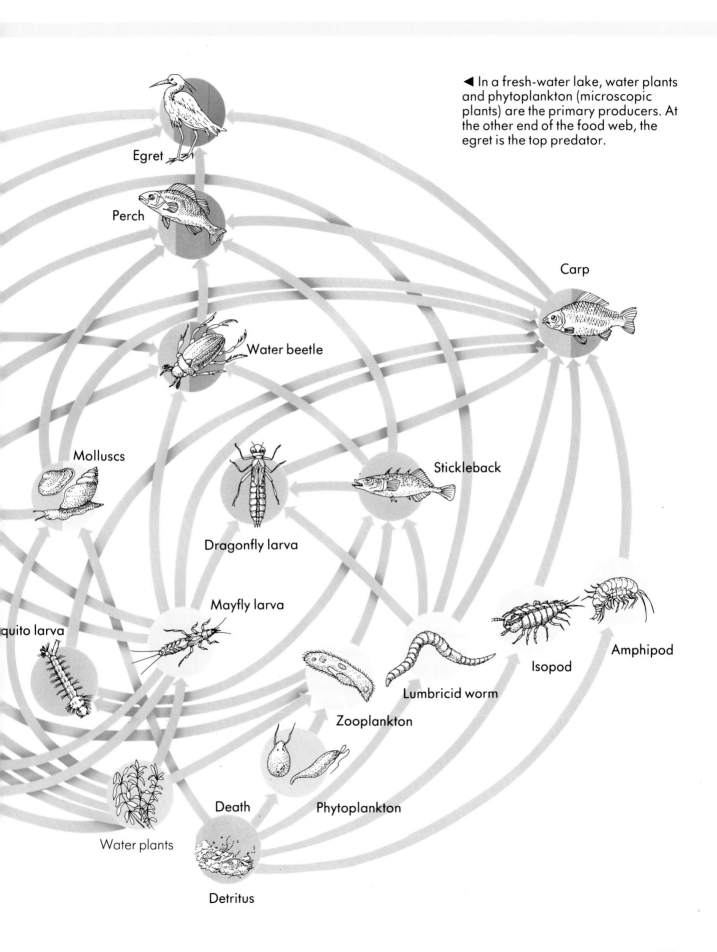

◀ In a fresh-water lake, water plants and phytoplankton (microscopic plants) are the primary producers. At the other end of the food web, the egret is the top predator.

Egret

Perch

Carp

Water beetle

Molluscs

Dragonfly larva

Stickleback

Mayfly larva

Amphipod

Isopod

quito larva

Lumbricid worm

Zooplankton

Death

Phytoplankton

Water plants

Detritus

In the rain forest

The rain forests of South America, Africa and South-east Asia are among the most complex ecosystems that exist. The hot, wet conditions have encouraged the development of an incredible variety of species. More than half the species on Earth may be found in these forests. This diversity of life is one of the main characteristics of these ecosystems.

The unvarying climate ensures a year-round supply of fruit and flowers. The variety of habitats and food sources provide countless opportunities for primary consumers. These in turn create niches for carnivores. In general, populations tend to be small, with many species existing in self-contained "island" communities.

A layered environment

The rain forests form a multi-layered, three-dimensional environment. At the top is the canopy, a mass of leaves over 30 m above the ground. The middle layer, the understorey, is fairly open, but is criss-crossed with creepers. The ground is largely free of vegetation, because little sunlight can penetrate through the dense overhead cover.

Animal life is concentrated in and around the canopy, attracted there by the limitless supplies of food. The understorey forms a kind of internal road network. The uncluttered space allows rapid movement by birds and monkeys. The ground is largely left to insects and the forest's few large mammals.

Animal life plays a vital role in maintaining the forest. Many species carry pollen between flowers as they feed. The seeds of some trees will not germinate unless they have passed through the intestine of a particular species of monkey. Using monkeys to scatter seeds far and wide is just one of the ways the forest preserves its enormous variety of plant species.

▶ Examples of most kinds of animal live in the rain forest. Snakes, such as the Boa constrictor (1), are generally found near the forest floor, although many are excellent tree climbers. Insects are found at all levels; the Red stainer bugs (2) feed on fruit that has fallen to the ground. The coati, a small mammal (3), hunts lizards and insects, mainly on the ground. Fruit-eating bats (4) are widespread, and the Howler monkey (5), is one of the largest mammals to inhabit the canopy. Its distinctive call can be heard echoing through the forest.

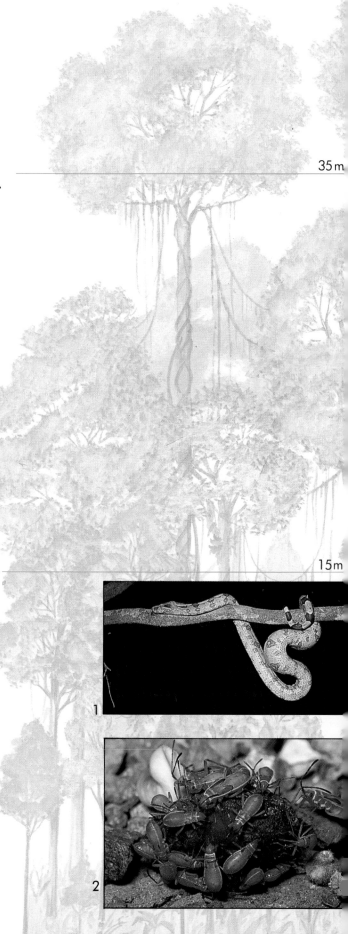

35 m

15 m

5

4

3

In the oceans

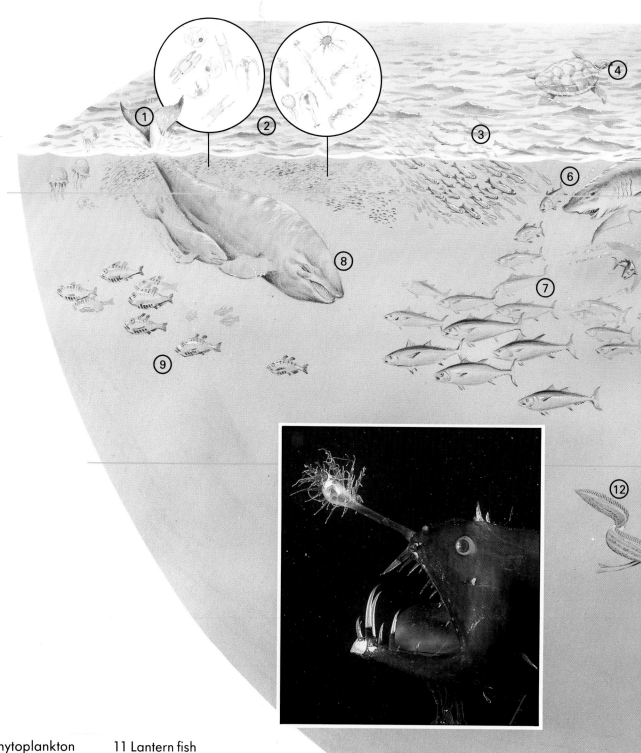

Key

1 Phytoplankton
2 Zooplankton
3 Anchovy
4 Green turtle
5 Dolphin
6 Shark
7 Bluefin tuna
8 Grey whale
9 Hatchet fish
10 Squid

11 Lantern fish
12 Oarfish
13 Giant squid
14 Deep-sea jellyfish
15 Skate
16 Brittle star
17 Deep-sea shrimp
18 Angler fish
19 Tripod fish
20 Sea cucumber

▲ This Angler fish is a deep-sea predator, and has luminous tentacles which attract smaller fish to eat in the dark waters of the ocean bottom. Food resources are scarce in the murky depths, and so most species rely on the scanty amounts of detritus (dead organic matter) that drift down from the upper layers. The Angler fish is in danger of attracting enemies with its lure as well as its prey.

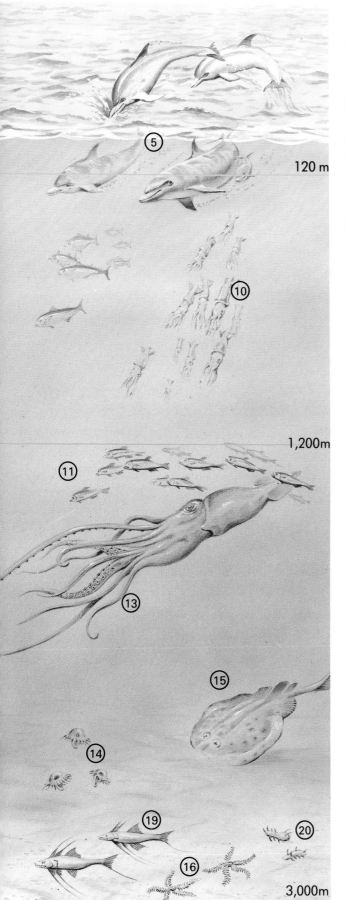

120 m

1,200m

3,000m

▲ The coral reef is a very rich habitat with thousands of species dependent on each other. At the bottom of the food web are the tiny plant plankton.

The oceans, which cover 70 per cent of the Earth's surface, contain a number of distinct ecosystems. The most productive areas are coral reefs and shallow coastal waters, where every available surface teems with life.

The ecosystem of the deep ocean falls into distinct layers. Near the surface live the microscopic plankton that are the basis of the food web. Most of the familiar species of fishes inhabit this layer, together with whales, dolphins and other aquatic mammals. Below 1,000 m, life is much scarcer because plants cannot live in the permanent darkness. Fish in this layer tend to be much smaller, although carnivores like the giant squid can grow to great size. There are also giant worms.

Plankton are the basis of life in the oceans. There are two types: microscopic plants (phytoplankton) and tiny animals (zooplankton). At the other end of the ocean's food web are the predators such as the shark one of the top predators.

207

Habitats at risk

Spot facts

• *200,000 sq km of rain forest are destroyed every year. If this continues, the rain forest could disappear completely within the next 50 years.*

• *11 million hectares of crop-growing land are lost each year because of soil erosion. An additional 7 million hectares of grassland are lost to the gradual process of desertification.*

• *Lake Volta in Ghana, West Africa, is the world's largest artificial lake. Formed by the Akosombo Dam, it now covers 8,500 sq km of drowned land.*

▶ This lizard is one of the lucky ones being rescued by nature conservationists as its habitat is being destroyed. It lived in the area flooded by the new reservoir created by the Itaipu Dam in Brazil. There is frequently an outcry at the destruction of natural habitats during land development, but rarely can the work be halted to save them.

Our own species, *Homo sapiens sapiens*, thrives on planet Earth. Human ingenuity has enabled our population to rise way beyond any natural limits, but only at a considerable cost to nature. Our method of food production, agriculture, is not a natural process. In one sense, we are the ultimate predators, because we consume entire habitats in the struggle to feed ourselves. Sometimes we succeed only in creating wasteland which cannot be used by wildlife or ourselves.

Mining, road building, and many other human activities also threaten the balance of nature. As a result, wildlife habitats around the world are now at risk. Only concerted action, by governments, organizations and individuals, can save them.

Nature and humankind

In 1960 the world's human population reached 3,000 million; today it is over 5,000 million and still rising. This increase in numbers places enormous pressure on the planet's resources, especially for food. Each night, one in seven human beings goes to sleep hungry. Until very recently, growing more food brought us into direct conflict with natural habitats. It meant cutting down more trees and clearing more land for the plough.

For centuries, agriculture has been transforming the landscape. We first started clearing the forests over 5,000 years ago to make room for flocks and herds. Agriculture brings us into direct conflict with nature. When farming, we are imposing our own food web, with human beings at the top, on to natural ecosystems. Anything that threatens our food supplies threatens our lives. We compete with wildlife for space in which we need to grow food.

Food is not the only burden that human beings place on the environment. We are the only species with a completely unnatural lifestyle. People also need water, clothing, fuel and shelter. Half the world's timber production is burned as firewood, and the cheapest methods of extracting minerals are usually the most destructive.

Physical space is also required because people have to live somewhere, and increasingly they are moving into the wild. Sometimes the movement is bold and dramatic, as when millions of settlers move to new homes on uncleared land. But in general, there has been a slow and steady invasion of natural habitats.

▼ Human population is increasing rapidly. Every hour, 8,000 people are born. Most of the increase is taking place in tropical countries where the climate makes large-scale agriculture extremely difficult.

Using the land

The environment and landscapes that we know today have only appeared recently in our planet's history. They are largely the result of dramatic changes in the Earth's climate.

Many of the basic landforms were sculpted by glaciers during the last Ice Age. About 30,000 years ago, huge sheets of ice covered much of Europe and North America, and areas of forest were very much smaller. The ice-sheets began to retreat about 20,000 years ago. As they did, they uncovered huge areas of land enabling the forests to increase in size. This process is still continuing, and in Alaska, in the United States, there are spruce trees growing where there were glaciers only 200 years ago.

Ice ages are not the only form of landscaping that is beyond human control. Earthquakes, volcanoes, violent storms, even meteorites from outer space can also change the shape of the world we inhabit.

World land use

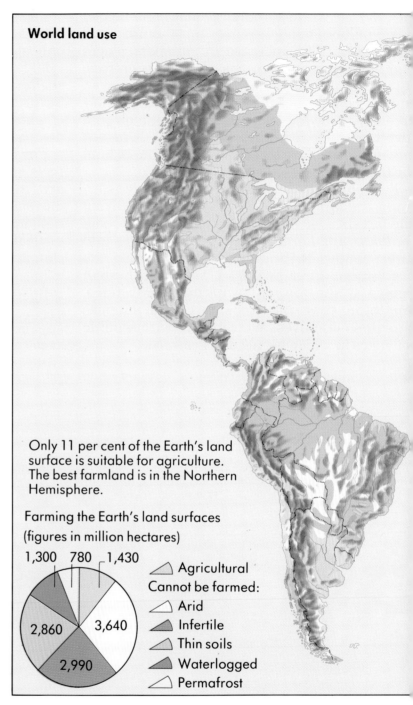

Only 11 per cent of the Earth's land surface is suitable for agriculture. The best farmland is in the Northern Hemisphere.

Farming the Earth's land surfaces (figures in million hectares)

1,300   780   1,430
2,860   3,640
2,990

△ Agricultural
Cannot be farmed:
△ Arid
△ Infertile
△ Thin soils
△ Waterlogged
△ Permafrost

Virtually all the land that is suitable for agriculture is now farmed. In the poorer countries, many people have to grow food on second-rate land that cannot sustain long-term usage. Crops become poorer every year.

Increased standards of living in the richer countries have also created greater demand for non-essentials. The result is that natural habitats are disappearing all over the world.

The northern limits of agriculture are being extended by the introduction of hybrid crops that mature during short summers. Grain fields now extend into the coniferous forest zone.

The world's temperate grasslands have long since been transformed by livestock and the plough. In the United States, there are only a few hectares of untouched prairie, which are now carefully preserved.

Arable land
Permanent pasture
Other grazing land
Forest
Land not farmed

In Africa, the Sahara Desert is slowly expanding southwards. The fragile ecosystem at the ringe of the desert exists in precarious balance, with uncertain rainfall. Under the weight of human numbers, and overgrazing by their flocks, the ecosystem is breaking down completely. Without plants to hold the land together, there is no barrier to the approaching sand. Not very long ago, trees covered the whole area.

Machinery greatly multiplies the destructive effects of agriculture. In Europe and North America most of the natural forests were cut down long ago. A second wave of habitat destruction is now threatening those woods, hedgerows and ponds which were unaffected by traditional farming methods. These refuges for wildlife are now being removed to create the huge fields required by modern machines.

Laying waste

The most dramatic example of habitat loss is the destruction of the tropical rain forests. The rain forests are disappearing at a rate of 2 per cent per year. Ranching, plantations, mining, logging and human settlement all demand their share of land. Within our lifetimes, the rain forests may disappear completely. The greatest threat to them is agriculture. In South and Central America the rain forests are being burned down to create vast cattle-ranches. Nearly all the meat produced is exported to be used in fast-food hamburger production.

This policy represents a very short-sighted use of resources which cannot be replaced. Heavy tropical rains will soon wash all the nutrients from the exposed soil. Experts predict that within 10 years, not even grass will grow where giant forest trees once stood. Once the grass dies, the soil itself will be washed away. Many scientists fear that the huge areas which have already been cleared will never regrow. The area could become a desert.

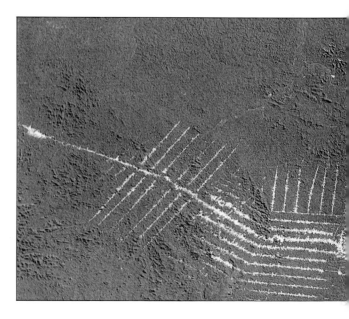

▲ The pattern of rain forest destruction in Brazil is clearly shown in this satellite image in which the vegetation shows up as red. Side roads branching off the main road give access to tree-cutting crews.

On a global scale, the rain forests represent huge masses of photosynthetic vegetation. They are responsible for recycling thousands of tonnes of carbon dioxide and oxygen every day. Without the rain forests, our atmosphere will gradually deteriorate.

Equally important is the rain forests' recycling of water. Over 75 per cent of the rain that falls on the forests is returned to the atmosphere through evaporation and plant respiration. This water may travel halfway around the globe before it falls as rain again. Perhaps a quarter of the world's human population depends on this water from the rain forests.

Preserving the rain forests is not just a matter of saving some exotic species; it might be a case of saving life on Earth.

► Land in Upper Egypt poisoned by excess salt (salination). Irrigating the desert is one of the chief causes of salination. As the irrigation water evaporates, it draws salt in the soil to the surface.

Unsuitable methods of agriculture can easily turn wild land into a wasteland. Natural grassland may include up to 40 different plant species. This varied mixture enables the ecosystem to flourish in regions of uncertain rainfall. There are enough drought resistant species to survive dry spells. When people plough up the grassland to create fields of a single crop, they upset the careful balance of nature. If that crop dies through drought, there is nothing to hold the soil together. It turns to dust, and may be blown or washed away.

This occurred on a large scale in the United States during the 1930s. Several years of low rainfall turned thousands of hectares of apparently good farmland into a Dust Bowl. Logging operations are also tremendously destructive, and the damage is not confined to the valuable hardwood of the rain forests. For example, it takes over 400 hectares of medium-sized coniferous trees to make a single edition of a Sunday newspaper.

◄ A landscape transformed by overgrazing in Australia. Large herds of livestock are doubly damaging. They eat all the available vegetation, and also trample plants into the ground, preventing regrowth.

Engineering the landscape

▲ Opencast china clay mine on Dartmoor, England. The local devastation caused by such mining operations is clearly apparent. In addition, mineral-laden dust may be carried great distances by the winds.

Increased human population, concentrated in rapidly expanding cities, has greatly increased demand for fuel and minerals. The Earth contains a huge range of mineral treasures. Fuels such as coal, oil and uranium, metals like aluminium and iron, and even the raw materials for concrete and glass, all have to be extracted from the ground.

Most of the mineral deposits that are readily accessible have already been exploited. The search for new supplies of fuel and metals is taking place in ever more remote areas, such as Siberia, Alaska and the Amazon Basin.

A new discovery usually spells disaster for the surrounding habitat. Humans and machines move in to tame the wilderness, felling trees and tearing up the land.

In the case of gas and oil, the greatest damage is caused by the heavy equipment used to establish the field, and with time the ecosystem can usually recover. Long-term damage, however, may be caused by pipelines carrying hot crude oil across landscapes frozen into permafrost, especially if pipes crack.

Opencast mining, employed where the mineral deposits lie just beneath the surface, is the most destructive form of exploitation. Coal, lignite, and many metal ores are commonly obtained in this manner. Huge, gaping holes are dug in to the ground, disfiguring the landscape for many years.

Mining deep underground, for coal or metals, creates the opposite problem. The surrounding area becomes dotted with unsightly slag heaps, artificial hills made of mud and crumbling stone. In the long term, both forms of mining are equally destructive because bulky minerals require roads and railways for transportation.

Roads generally mark the beginning of the end for natural habitats. They open up the land for other forms of exploitation. First the area at the roadside is cleared for firewood, then for recreation, and finally for towns and cities.

Diverting the waters

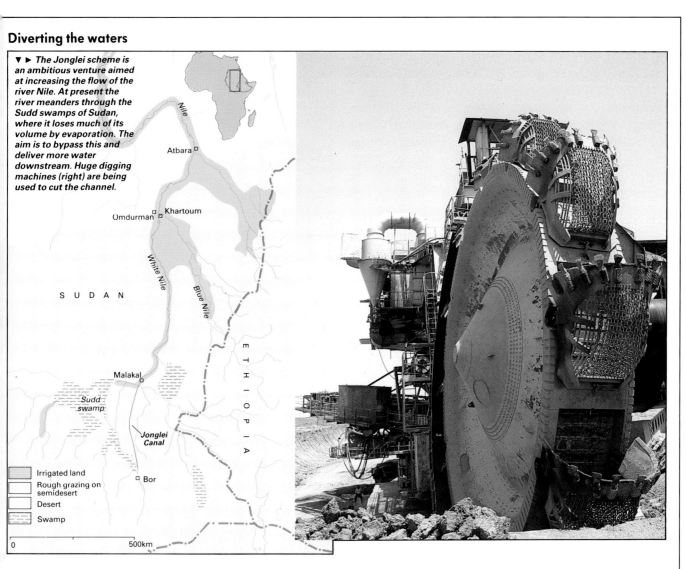

▼ ► The Jonglei scheme is an ambitious venture aimed at increasing the flow of the river Nile. At present the river meanders through the Sudd swamps of Sudan, where it loses much of its volume by evaporation. The aim is to bypass this and deliver more water downstream. Huge digging machines (right) are being used to cut the channel.

Nile

Atbara □

Umdurman □ □ Khartoum

White Nile

Blue Nile

S U D A N

E T H I O P I A

Malakal □

Sudd swamp

Jonglei Canal

□ Bor

Irrigated land

Rough grazing on semidesert

Desert

Swamp

0   500km

▲ The Kariba Dam across the River Zambesi in East Africa is one of the world's largest, and was built to provide hydroelectric power.

▲ Throughout history, the most ambitious engineering schemes have been concerned with water, or hydro-engineering. There have been dams to hold back irrigation water, aqueducts to carry drinking water, and canals for transport and drainage. Today, some of the biggest machines ever built are digging the Jonglei Canal in north-east Africa. The aim is to increase the amount of water in the River Nile by diverting water from the Sudd swamps in Sudan where much of it evaporates. Huge digging machines are being used to cut a deep channel for the water. Some experts believe that the project will provide much-needed irrigation water for farmland downstream. Others fear that the canal will turn the Sudd swamps into desert.

A similar scheme in southern Russia has caused the Aral Sea, once the world's fourth largest inland sea, to lose 69 per cent of its water in less than 30 years. The shrinking of the Aral Sea has caused an ecological disaster affecting 30 million people.

Preserving the habitat

There are many organizations, at a local, national and international level, devoted to preserving the environment. Together, they form the "Green movement". In several countries there are Green political parties.

Whenever a habitat is threatened by industry or development, the Green movement takes the side of the environment. These groups already enjoy considerable public support, but their most important work is education. Unless people understand how threats to the environment affect them, they cannot appreciate the importance of preserving habitats. Every hectare that we save today is one that will be enjoyed by our grandchildren.

▼ Natural grassland, with its springtime carpet of wild flowers, is becoming rare in Europe and North America. Unless action is taken to preserve the few remaining areas, the flowers will disappear under the plough.

Many farmers are sympathetic to the Green movement, despite the pressure to grow the maximum amount of food. Some make an effort to preserve small areas of woodland and hedgerow, and are careful not to spray these areas with pesticides. A few have turned their backs on modern methods and use no chemicals at all, either in fertilizers or pesticides.

Farming without chemicals is known as organic farming. In many countries, organic produce has become very popular. Because no chemicals are used to grow the food, there are none to enter the human food chain. In addition, by choosing organic produce, people are also choosing to save the countryside.

In the developing countries, action is finally being taken to save the rain forests. In Africa, the Ivory Coast has recently banned all timber exports. In Central America, the government of Panama has made it illegal to cut down any tree more than five years old. Such actions show great determination, because timber exports have been a major source of income.

It is difficult to persuade poor people in developing countries of the importance of saving the forests. They need more food and want to use the land to grow it on. But saving habitats is not a luxury, it is a necessity. What is at stake is not just our enjoyment of nature, but the future of life on Earth.

◀ Wicken Fen Nature Reserve in England. Wetlands, such as marshes and bogs, are extremely rich in rare plant and animal species. Unfortunately, they also make excellent farmland after they have been drained.

▼ Yellowstone National Park in the United States, famous for its Grizzly bears and spectacular geysers. Most countries have established National Parks where wildlife is protected and development is prohibited.

Species at risk

When a species becomes extinct, it disappears for ever. Extinction is a natural part of evolution, but it is usually a slow process. Human beings speed things up. Some animals that are known from photographs taken during the last century are now extinct. Today, a huge international trade in animals and animal products is consuming the planet's wildlife. We are also destroying the habitats in which they live. Directly or indirectly, human activity now threatens many familiar species. Yet it is only through our intervention that many endangered species have any chance of survival at all. On the other hand, some wildlife species thrive in human company, but these are often the least welcome.

▶ The Golden lion tamarin is a small monkey and now very rare in the wild. It lives only in forest remnants in eastern Brazil. Hundreds have been exported for zoos and for the pet trade. This is now illegal, but still continues. Their habitat is almost completely destroyed by development for the tourist trade.

Animal products

Human beings have an insatiable appetite for animal protein. In China, where almost everything is considered edible, the wildlife is steadily being eaten to extinction. The world's oceans are under similar threat. Overfishing can cause dramatic changes in the size of marine populations. In 1937 California landed 750,000 tonnes of sardines; by 1957 this had fallen to a mere 17 tonnes. Such a rapid decrease in numbers creates a gap in natural food chains that can affect other species.

At the beginning of the last century, there were perhaps 40 million bison roaming the North American prairies. By 1900, only 500 remained. The vast majority of those slaughtered were left to rot. The hunters only wanted the hides for leather.

Many other species have been hunted to the point of extinction for their skins and furs. Countless beavers, otters, seals, minks, bears, leopards and other big cats have been sacrificed in the name of fashion and status. Human fancies have created an international trade in animal products: exotic feathers for hats, turtle shells for combs, and recently the craze for rare and exotic pets.

Easy money is the motive. The poachers become rich, while the world becomes poorer. Several species of rhinoceros have disappeared because their horns were thought to have magic powers. Today, the African elephant is being wiped out for the sake of its tusks. The ivory is considered valuable because it can be carved into souvenirs for tourists.

◄ These African zebra skins may be destined to upholster prestige furniture in Europe or North America. Most of the demand for furs and skins comes from the developed countries which have destroyed nearly all their own wildlife. Animal products are a valuable export commodity for many developing countries.

▼ These ivory carvings are made from the tusks of the African elephant. The population of African elephant is dropping very quickly, largely because of ivory poaching. Poaching gangs operate in game reserves despite armed patrols. These carvings were seized by customs as ivory exports are illegal.

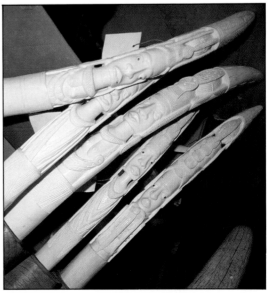

The march of extinction

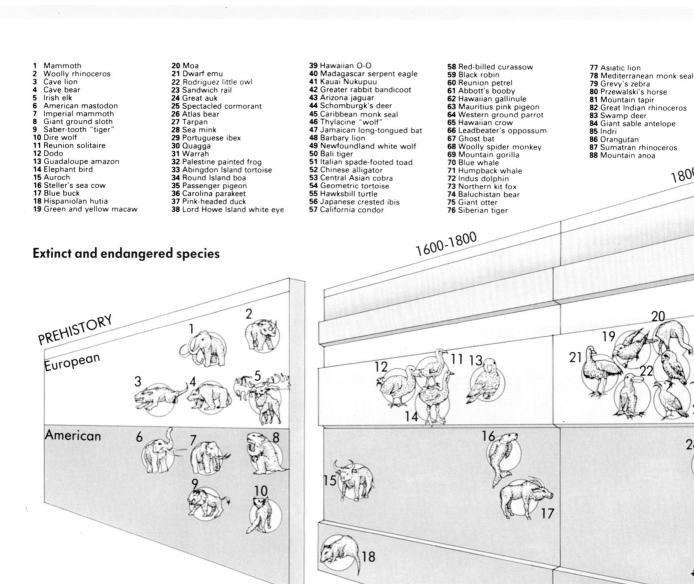

1 Mammoth
2 Woolly rhinoceros
3 Cave lion
4 Cave bear
5 Irish elk
6 American mastodon
7 Imperial mammoth
8 Giant ground sloth
9 Saber-tooth "tiger"
10 Dire wolf
11 Reunion solitaire
12 Dodo
13 Guadaloupe amazon
14 Elephant bird
15 Auroch
16 Steller's sea cow
17 Blue buck
18 Hispaniolan hutia
19 Green and yellow macaw

20 Moa
21 Dwarf emu
22 Rodriguez little owl
23 Sandwich rail
24 Great auk
25 Spectacled cormorant
26 Atlas bear
27 Tarpan
28 Sea mink
29 Portuguese ibex
30 Quagga
31 Warrah
32 Palestine painted frog
33 Abingdon Island tortoise
34 Round Island boa
35 Passenger pigeon
36 Carolina parakeet
37 Pink-headed duck
38 Lord Howe Island white eye

39 Hawaiian O-O
40 Madagascar serpent eagle
41 Kauai Nukupuu
42 Greater rabbit bandicoot
43 Arizona jaguar
44 Schomburgk's deer
45 Caribbean monk seal
46 Thylacine "wolf"
47 Jamaican long-tongued bat
48 Barbary lion
49 Newfoundland white wolf
50 Bali tiger
51 Italian spade-footed toad
52 Chinese alligator
53 Central Asian cobra
54 Geometric tortoise
55 Hawksbill turtle
56 Japanese crested ibis
57 California condor

58 Red-billed curassow
59 Black robin
60 Reunion petrel
61 Abbott's booby
62 Hawaiian gallinule
63 Mauritius pink pigeon
64 Western ground parrot
65 Hawaiian crow
66 Leadbeater's oppossum
67 Ghost bat
68 Woolly spider monkey
69 Mountain gorilla
70 Blue whale
71 Humpback whale
72 Indus dolphin
73 Northern kit fox
74 Baluchistan bear
75 Giant otter
76 Siberian tiger

77 Asiatic lion
78 Mediterranean monk seal
79 Grevy's zebra
80 Przewalski's horse
81 Mountain tapir
82 Great Indian rhinoceros
83 Swamp deer
84 Giant sable antelope
85 Indri
86 Orangutan
87 Sumatran rhinoceros
88 Mountain anoa

Extinct and endangered species

Recently-discovered evidence suggests that early humans played an important role in the extinction of some prehistoric animals. During the last 500 years, people have certainly left their mark on the world's wildlife. Dozens of species known to our ancestors have now vanished. Only pictures and descriptions remain. Species that live in islands are more vulnerable than those on the mainland.

Perhaps the most frightening example is the North American Passenger pigeon. During the early 1800s, a single flock was estimated to contain over 2,000 million individuals. Through a combination of habitat destruction and hunting for food and sport, the bird had disappeared from the wild by the end of the century. The last ever Passenger pigeon died in a zoo in 1914.

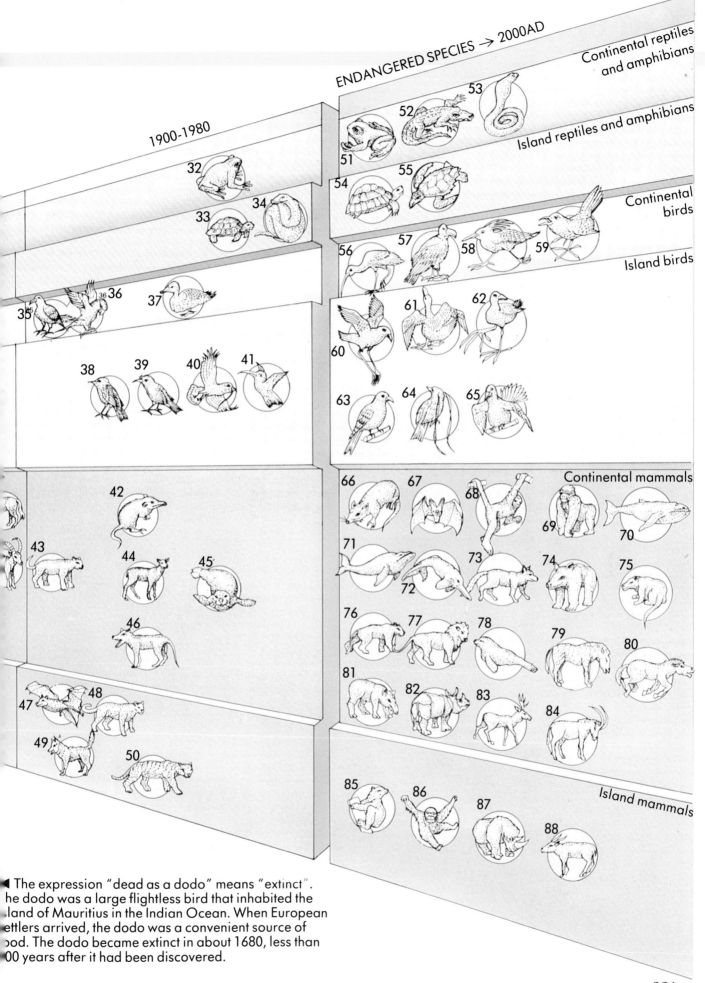

Continental reptiles and amphibians

Island reptiles and amphibians

Continental birds

Island birds

Continental mammals

Island mammals

1900-1980

◄ The expression "dead as a dodo" means "extinct". The dodo was a large flightless bird that inhabited the island of Mauritius in the Indian Ocean. When European settlers arrived, the dodo was a convenient source of food. The dodo became extinct in about 1680, less than 00 years after it had been discovered.

221

On the edge

Today, some of the world's most familiar animal species stand at the edge of extinction. Over 6,000 species are officially acknowledged to be endangered, and the list is far from complete. Large animals are especially at risk from habitat destruction. Their natural populations tend to be quite small, because each individual requires proportionately more territory.

The African elephant is the world's largest land animal. During the last ten years, its numbers have been halved from 1.3 million to 600,000, by ivory poachers. Its smaller cousin, the Asiatic or Indian elephant, is reduced to about 30,000 individuals living in the wild.

Survival at sea

In the oceans, the aquatic mammals are perhaps most at risk. Twelve species of whale are endangered, and the Grey whale has only been saved from extinction by 50 years of international protection. The one remaining species of dugong, or sea-cow, is still being hunted in Indonesia. In the Mediterranean, the Monk seal population is now less than 1,000.

Nor are human beings especially careful about their own close relatives. There are at least 50 species of primate in danger. The gorillas and the orang-utans now exist only in very small numbers. Declining chimpanzee populations are causing concern throughout Africa. Many small animal species, such as the Golden marmoset, are now seriously threatened by the growing demand for exotic pets.

Many extinctions have been caused by humans introducing animal predators. For the last 50 years, no offspring of one species of giant Galapagos tortoise have survived. The young are eaten by black rats which arrived on the Galapagos Islands with human settlers. Other unique island species are also under threat. In Madagascar, populations of some rare primates such as aye-ayes and lemurs number less than 100 individuals.

▶ The Giant panda is found only in a small area of southern China. There are about 200 pandas left in the wild, and they very rarely breed in captivity. When the last wild panda dies, extinction will surely follow.

▲ The shy and gentle Mountain gorilla is one of the world's rarest animals. Its habitat in West Africa is disappearing under the pressure of human cultivation. Tourists who are drawn to these remote areas to see wild gorillas are increasingly damaging the habitat.

◄ The Southern Right whale is one of the world's largest living animals. Formerly killed in large numbers, they are now "commercially extinct". This means that it is not worth the expense of hunting the last few. This has extended the species' life a little, but there may not be enough individuals left for the population to recover.

223

Saving species

◀ Przewalski's horse now exists only in captivity, but in sizeable numbers. Zoologists would like to release some back into Central Asia, but they are concerned that it has become adapted to life in zoos.

▶ The White rhinoceros, like all rhinoceros species, is seriously endangered. In 1979 Project Rhino was launched to coordinate attempts to save them. Poachers are still a deadly threat to rhinoceroses.

▼ The Arabian oryx became extinct in the wild in 1972. Fortunately, Operation Oryx had already established a small herd in an American zoo, which bred success-fully. During the late 1970s the oryx was released back to the wild.

Perhaps the most important thing we can do for endangered species is to recognize their plight. Publicity from concerned governments, and organizations such as the World Wide Fund for Nature (WWF), have done a great deal to raise public awareness. Intervening to solve the problem, however, is very difficult.

Zoos provide endangered species with a mixed blessing. Although care may be given them in captivity, the methods of capture are not always so gentle. The normal way of "taking" a young mammal is to first kill its mother. With tropical birds, as many as 50 may be killed for every live specimen taken.

Game reserves, which protect animals and their natural habitat, are much more humane. But the habitat may be too inaccessible, and there is the problem of providing armed guards. Many poachers will start a gun battle with the game wardens rather than be caught.

If the species will breed in zoos, there is hope that a population may eventually be released back to the wild. Some species, however, do not breed in captivity. In a few cases, entire wild populations have been carefully collected, and then transported to a less threatened habitat. Sometimes they have been released on to a game reserve, but for others a remote island has offered the only chance of safety. But there are fewer and fewer wild places left where animals can be safe from our exploitation of the land.

In many parts of the world, governments and international organizations have successfully cooperated to protect endangered wildlife. A series of projects has given several species a fighting chance of survival. At the other end of the scale, important work is also done by small local groups devoted to saving a single plant or insect species. One of the most important lessons of ecology is that every species has its role to play in the balance of nature, and therefore every species is important.

Project Tiger

The Bali and Caspian tigers are now extinct. The Siberian and Javan tigers are unlikely to survive the century outside zoos. The Bengal tiger is the only species that is likely to survive in the wild. Project Tiger is the master-plan for saving the Bengal tigers.

The Indian government, with support from the WWF, has established nine tiger reserves. Trained rangers and armed guards protect the animals against poachers and forest fires. Some animals are fitted with electronic collars to monitor their movements.

Thriving with people

Representatives of several animal groups – mammals, birds and insects – have adapted to human existence so well that they have become pests. Most of them have completely abandoned a wild existence, and have become parasites on human food webs. From the time people began sailing the world, they travelled around the globe in ships' cargoes. These species now live near people all over the world.

There are over 3,500 known species of cockroach. A few of these have become such close companions of people that "roaches" are now almost universally despised. Able to eat virtually anything, and resistant to insecticides, cockroaches spoil human food and spread disease. The Oriental and German cockroaches have become a familiar sight in millions of homes around the world.

The House mouse originated in Central Asia but has spread around the world. The species has found a niche in the nooks and crannies of human habitations. Maintaining a cautious and nocturnal existence, mice scavenge food from human stores and waste. Its larger cousins, the Brown and Black rats, are far more destructive. Both rat and mice populations can undergo very rapid growth when conditions are good. These pests can consume huge quantities of stored food in homes and warehouses.

Larger woodland mammals are slowly coming to terms with humans, and the process is speeded up by the growth of suburbs. Racoons or foxes are quite a common sight in country gardens, and both species are now adapting to life in the city. In parts of Alaska, Polar bears are known to raid dustbins.

Common racoon

Black-headed gull

Herring gull

House mouse

Norway rat

Feral pigeons

Common starling

▲ ▼ These species of animals are all found near human dwellings and are often thought of as pests. Some carry infectious diseases. The plant species all grow in waste ground in cities. Although regarded as weeds, they are pioneers colonizing barren ground, and support many different kinds of insects.

Rosebay willowherb

House sparrow

Small white butterfly

Buddleia

Shepherd's purse

Narrow-leaved plantain

Oriental cockroach

Several species of seagull, especially the Herring gull and Black-headed gull, have partially adapted to life on land. Many of them have left the seashore, and they now pick over urban rubbish dumps. Even far inland, they are a familiar sight flocking around the plough.

The most common urban birds are sparrows, starlings and pigeons. In fact, we have become so attached to these species that they have often been deliberately introduced to new cities. Sparrows nest in old houses, trees, parks and gardens. Pigeons find that tall modern buildings provide excellent roosting sites. They are also common, if unwelcome, in railway stations.

Starlings also roost in cities, but unlike their cousins they are commuters. Pigeons and sparrows stay inside the city, scavenging where they can. Starlings fly into the countryside each day to feed on crops up to 30 km away. At night they return to the city to sleep.

As well as parks and gardens, waste ground also provides an urban habitat for wildlife. Among the colonizing plants, many of them considered weeds, animals live in secret communities. Buddleia, frequently seen on derelict land in cities, is very attractive to butterflies. Even in the most unnatural conditions, wildlife will flourish.

Pollution

Spot facts

● *110 million tonnes a year of sulphur dioxide (a major cause of acid rain) are released into the atmosphere.*

● *Over 8,000 different pollutants have been identified in water taken from lakes.*

● *Cleaning up after the nuclear accident at Three Mile Island in the USA will cost over $1,000 million.*

● *Pollution travels. Acidic chemicals can be carried 2,000 km by the wind before falling to Earth as acid rain.*

● *800 years ago King Edward I of England imposed a death penalty on anyone found burning coal because it created noxious fumes.*

▶ This Western grebe is being cleaned of the sticky crude oil which prevents it flying and feeding. The oil spilled accidentally into the San Francisco Bay, USA, from a Standard Oil tanker. Emergency bird-cleaning teams can help rescue some seabirds after a large disaster such as this. Pollution of the sea, lakes and rivers causes untold damage to wildlife.

Human beings are the only species that poison their own habitat. If we think of our planet as a spaceship, then we are the crew and all the other species are passengers. In general, we behave like hooligans, strewing the communal living-quarters with rubbish. Earth's atmosphere forms a protective cocoon around our spaceship, which uses it to store and recycle essential elements such as carbon and oxygen. Humans fill the air with pollution that turns the rain acidic, and threatens to alter the climate.

Pollution from fossil and nuclear fuels, from industry and agriculture, are slowly poisoning the land, air and water that we share with all living things.

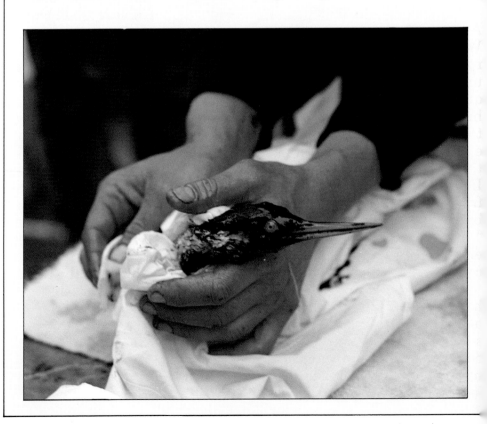

Poisoning the world

▲ Workers in protective clothing and masks clearing up after an explosion at a pesticide factory in Seveso, Italy. The ingredients of pesticides include some of the most poisonous chemicals we know.

◄ Mountains of solid waste disfigure the landscape. Little grows here. Metal and paper will eventually rust and rot, but some of the plastic will last for many years before it eventually decomposes.

All living organisms produce waste products. With all species except humans, these are safely recycled by the ecosphere. We alone produce unnatural waste, and we produce it in huge quantities. Atmospheric pollution began when human beings first discovered fire, but for thousands of years it had little impact on the environment. Serious problems started during the 1800s with the coal-burning factories of the Industrial Revolution. The smoke belching from their chimneys turned buildings and trees black with soot. Today, factory smoke is much "cleaner". The black soot has largely been eliminated, but invisible combustion products continue to pour into the air.

In the 1900s, huge numbers of motor vehicles have greatly increased the burden on the atmosphere. Vehicle exhaust fumes contain several harmful substances. These include carbon monoxide, a highly poisonous gas, nitrogen oxide, which produces acid rain; and lead, which is deadly to most forms of life. The volume of motor traffic is steadily increasing.

Industry is responsible for a wide range of other pollutants. Many of the chemicals used in industry are poisonous, and others are known to cause cancer or deformities in babies. In many countries, especially in recent years, industry is very careful, but accidents do happen. A single chemical spillage can cause a river to become lifeless for years. Some companies have disposed of their waste chemicals by dumping them on empty land. This land is now completely uninhabitable.

Most modern farmers use large quantities of powerful chemicals to kill weeds and insects. Even in the dilute concentrations used in spraying, these can damage wildlife.

Some of these chemicals tend to accumulate in animals' bodies. The insecticide DDT is a notorious example. When such a chemical enters the food chain, it gets passed along. The higher up the food chain an animal is, the more chemicals it absorbs. This process is known as bio-concentration. Predators, such as birds of prey, are especially at risk.

Polluting the air

Each year millions of tonnes of pollution from power stations, factories and motor vehicles enters the atmosphere. Most countries have agreed to reduce the pollution caused by industry. Some now require motor vehicles to be fitted with devices to filter out pollutants, or to use cleaner types of petrol.

Despite these measures, the effects of atmospheric pollution are often felt below. Acid rain is created when the products of burning coal and oil, combustion products, become combined with water in the atmosphere to form acids. Falling back to Earth as rain, these acids directly attack trees and plants. Acidic water also accumulates in rivers and lakes, killing fish and other aquatic life. The acid destroys the natural chemical balance.

In Britain, over 60 per cent of the forests have been harmed by acid rain, and large areas of North America are similarly affected. The winds often take acid rain to neighbouring countries. Southern Norway, for example, has few heavy factories, yet 80 per cent of its lakes are now devoid of life or on the critical list.

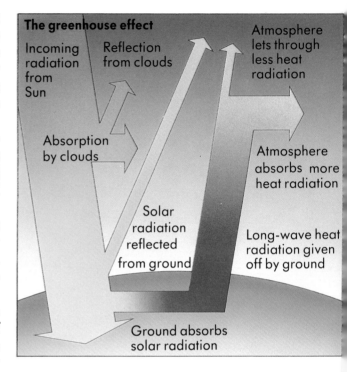

The greenhouse effect

Incoming radiation from Sun

Reflection from clouds

Atmosphere lets through less heat radiation

Absorption by clouds

Solar radiation reflected from ground

Atmosphere absorbs more heat radiation

Long-wave heat radiation given off by ground

Ground absorbs solar radiation

▲ Carbon dioxide has formed a blanket around the planet. Heat energy from the Sun is trapped by the atmosphere, which is slowly warming up.

Bhopal

Accidental air pollution by industrial chemicals can have disastrous local effects.

On 3 December 1984, there was an accident at a pesticide factory in the town of Bhopal, India. An explosion released 30 tonnes of methyl isocyanate, a highly toxic gas, into the air. Over 200,000 people in the vicinity were exposed to the gas. About 2,500 people were killed immediately, as many as 10,000 may have died subsequently, and at least 20,000 were disabled for life.

The Bhopal incident drew attention to one of the most important pollution issues. The rich countries of Europe and North America can afford to be concerned about pollution. They have well-developed industrial economies, and can impose high standards of safety and cleanliness. Poorer countries in the developing world often cannot afford to take these measures, even though they may wish to.

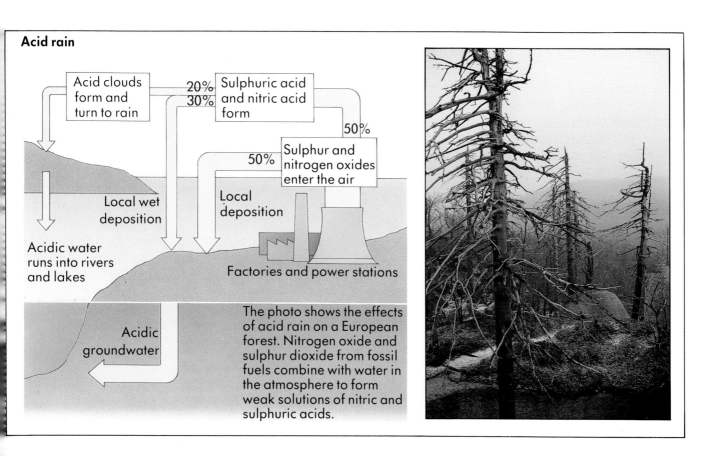

Acid rain

Acid clouds form and turn to rain

20% / 30% → Sulphuric acid and nitric acid form

50%

50% → Sulphur and nitrogen oxides enter the air

Local wet deposition

Local deposition

Acidic water runs into rivers and lakes

Factories and power stations

Acidic groundwater

The photo shows the effects of acid rain on a European forest. Nitrogen oxide and sulphur dioxide from fossil fuels combine with water in the atmosphere to form weak solutions of nitric and sulphuric acids.

Human activity is producing far more carbon dioxide gas than the oceans and forests can recycle. On a global scale, increased amounts of carbon dioxide in the atmosphere threaten to change the climate of the whole planet. As the gas accumulates in the upper atmosphere it creates an insulating layer around the planet. Scientists believe that average temperatures on Earth might rise by 4°C. The ice caps could melt and many countries be flooded by rising seas.

This problem is often called the greenhouse effect. Carbon dioxide from burning fuel is the main culprit, but other gases are also involved. The propellant gases in many aerosols, chlorofluorocarbons (CFCs) are especially damaging to the atmosphere. As well as adding to the greenhouse effect, CFCs are also dissolving the Earth's ozone layer. Without ozone to act as a filter, the Sun's rays can be harmful to living things.

◄ Smog over Mexico City. Fumes from the burning of fossil fuels can become trapped near the ground by a layer of air. The action of sunlight on the trapped fumes produces a poisonous smog.

Nuclear hazards

Radioactivity occurs naturally; many rocks are slightly radioactive. As life on Earth evolved, it adapted to this background radiation. Humans first discovered the secrets of atomic and nuclear power in the 1940s. Since then, pollution from unnatural radioactivity has become a potential health hazard. There has been considerable debate about what constitutes a safe dose of radiation. Today, most scientists agree that any exposure to unnatural radiation is dangerous, and a risk not worth taking.

The world's first exposure to radiation pollution followed the use and testing of atomic weapons. Nearly 1,500 bombs have been detonated, most of them in the atmosphere, although the earliest tests were made in deserts and on remote islands. These explosions have scattered tonnes of radioactive fall-out around the world. Some of the fall-out loses its radiation within a short time. But some of the radioactive substances, or radioisotopes, remain dangerous for many years.

These isotopes enter food chains via plants, and can accumulate in animal bodies and be passed along food chains. In the past there was considerable concern about the amount of the isotope strontium-90 in cows' milk. International agreement has now almost put an end to atomic tests.

Today, the use of nuclear power to generate electricity is the main focus of concern. Although nuclear power stations are designed to be as safe as possible, there is always some leakage of radiation during operation. There is also the problem of what to do with the radioactive fuel after use. Additionally, there is always the possibility of an accident.

▼ The nuclear power station at Three Mile Island, Pennsylvania, USA, has been shut down since the accident there on 28 March 1979. Mechanical failure caused the reactor to overheat, which might have led to a far greater catastrophe. After 18 hours, engineers were able to bring the reactor under control and so averted the danger of a deadly explosion.

There have been two major accidents at nuclear power stations. In 1979, a pump failed at the Three Mile Island installation in the United States. The operators managed to prevent an explosion, but the power station was wrecked and a considerable amount of radiation was released into the atmosphere.

In 1986, their Russian counterparts were not so successful, and the Chernobyl power station exploded. The explosion produced a cloud of fall-out that spread over 75 per cent of Europe. Even in Sweden, thousands of kilometres from Chernobyl, livestock had to be kept under cover to protect it from the fall-out. Experts predict that at least 1,000 people in Western Europe will eventually die as a result of eating food contaminated by the fall-out from Chernobyl.

These two accidents have caused many people, and some governments, to have second thoughts about nuclear power. There will always be a risk, and the next accident might be much worse.

▲ Checking vegetables with a Geiger counter, which measures radioactivity. Meat and milk products gave the most concern after Chernobyl. In Britain, a ban was put on the sale of sheep from areas that received the heaviest fall-out. Over 500 farms were still affected a year later.

◄ When the nuclear reactor at Chernobyl exploded in 1986, it blew apart the entire building and started a fire. Some of the first casualties were firemen who received fatal doses of radiation while bravely fighting the blaze. At least 200 Russians have so far died as a result of the explosion, and many towns and villages have been permanently evacuated.

233

Polluting the waters

All the waters of the Earth are bombarded with pollutants. Rain and rivers wash our waste into lakes and seas where it slowly accumulates. Natural processes can remove some, but by no means all, of the pollutants that household and industrial waste contains.

In Europe, the River Rhine alone carries over 300,000 tonnes of waste into the North Sea each year. Most of this is permitted waste: sewage and waste chemicals that most scientists believe the sea can safely absorb. But a series of accidents have also spilled many tonnes of highly poisonous substances.

Cities and industry are not the only cause of water pollution. The artificial fertilizers that many farmers depend on can also pollute water supplies. The phosphates and nitrates contained in artificial fertilizer are easily washed out of the soil by rain. These nutrient chemicals accumulate in rivers and lakes, where they destroy the natural balance of nutrients.

▲ The first oil tanker disaster to affect Britain was the wreck of the *Torrey Canyon* in 1967. After other measures had failed, the government had the wreck bombed, to prevent further pollution.

▲ (Inset) In 1978 an oil spillage from the tanker *Amoco Cadiz* devastated part of the French coastline. A national emergency was declared, and cleaning up operations lasted for months.

◄ Several countries continue to pump chemical wastes and untreated sewage into the sea. Many beaches have been declared unfit for swimming because of the potential health risks from this pollution.

The increased concentrations of nutrients stimulate algae (microscopic waterplants) to overgrow. When the nutrient resouces are all used up, the algae die. As they decompose, all the oxygen in the water is used up and the environment becomes lifeless.

This process, which is known as eutrophication, usually occurs in small bodies of water. Recently, however, it has been blamed for the great masses of dead algae that have been observed in some of the world's smaller seas.

Enclosed seas, such as the Mediterranean Sea and the Baltic Sea, are particularly at risk from water pollution. Their coasts are lined with industry, and they have only narrow outlets into the main oceans. Pollution therefore becomes more concentrated, and the effects on marine life are much greater. These seas may soon become completely dead.

Oil is probably Earth's most valuable and useful commodity, and hundreds of millions of tonnes are carried by tanker each year. Inevitably there have been accidents, and large quantities of crude oil have spilled into the sea.

Crude oil floats on the surface, and decomposes in a few weeks under the action of seawater and sunlight. When the spillage occurs close to land, and oil is washed ashore, a few weeks is far too long. The oil coats everything and has a disastrous effect on the local ecosystem. Seabirds and aquatic mammals sink and drown, shellfish are smothered, and many fish are poisoned.

Water pollution has spread to the poles, and has now entered the Antarctic food chains. The choice is quite simple: either we clean up, and halt the steady increase in pollution, or our planet will slowly die.

Index

Picture Credits

b = bottom, t = top, c = centre, l = left, r = right
A Ardea. ANT Australasian Nature Transparencies, Australia. BCL Bruce Coleman Ltd. FLA Frank Lane Picture Agency. FSP Frank Spooner Pictures. GSF Geoscience Features. HL Hutchison Library. MN Maurice Nimmo. NCAR National Center for Atmospheric Research, USA. NHPA Natural History Photographic Agency. NSP National Science Photos. OSF Oxford Scientific Films. PEP Planet Earth Pictures. RHPL Robert Harding Picture Library. SAL Survival Anglia Ltd. SPL Science Photo Library. TCL Telegraph Colour Library.

10 Vautier de Nanxe. 11l PD Moehlman. 11r NASA. 13t SPL/J Finch. 13b Michael Holford. 17 Spacecharts. 18l AC Wattham. 18r ANT/Mark Wellard. 20 RHPL. 21 MN. 23 Bildarchiv Preussischer Kulturbesik. 24-25 NSP/MS Price. 26-27 FSP: 28 Dr J Shelton. 29 FSP. 30 Zefa. 31 SPL/Sinclair Stammers. 34l GSF. 34r FSP/Gamma. 35l MN. 35r Spacecharts. 36l GSF. 36r MN. 36-37 Zefa. 37t, 37c MN. 37b Rida. 38 SPL/Alex Bartel. 39tl, 39tr, 39bl, 39br GSF. 39bc Imitor. 40, 41r GSF. 42l Zefa. 42 Dr J Shelton/US Geological Survey. 46 Zefa. 47tl Dr J Shelton. 47tr GSF. 47br GR Roberts. 48 Zefa. 49 GSF. 51 ANT/A Jackson. 52 PEP/Warren Williams. 52 inset SCL. 53 RD Ballard/Woods Hole. 54 GSF. 56 RHPL. 57t NHPA/S Dalton. 57b SCL. 58 SPL/Sinclair Stammers. 59 RHPL/Walter Rawlings. 60 TCL. 60-61 SPL/G Garradd. 61t Hutchison Library. 61b SPL/Jerome Yeats. 62 Spacecharts. 63 Remote Source/Luke Hughes. 66t, 66b GSF. 67t Hutchison Library/B Regent. 67b SPL/NASA. 68 NASA. 70 Rex Features/Sipa Press. 71 Linda Gamlin. 72tl, 72tr NCAR. 72b TCL/ESA/Meteo Sat. 73 Rainbow/Hank Morgan. 74 SPA/NASA. 76 Zefa. 77l J Mackinnon. 77r Michael Rogden. 78l Zefa. 78r SAL/Jeff Foott. 79tl SAL/Alan Root. 79bl PEP/Ernest Neal. 79br South American Pictures/Tony Morrison. 80t Jacana/Fred Winner. 80b NHPA/P Warnett. 81tl Ardea/S Roberts. 81bl L La Rue Jr. 81r FLA/Mark Newman. 82-83 Leonard Lee Rue III. 84l Vautier de Nanxe. 84r Agence Nature. 85t Bruce Coleman/Hälle Flygare. 85c FLA/M Newman. 85b PEP/Rod Salm. 86 OSF/JAL Cooke. 87t, 87b MR Walter. 88 Dr R Goldring. 89 S Conway Morris. 90l Imitor. 90r SPL/Sinclair Stammers. 91 RHPL, Walter Rawlings. 92 Zefa/Colin Maher. 94 SPL/Sinclair Stammers. 95tl SPL/Dr Ann Smith. 95bl SPL/Prof N Alvarez. 95r GSF. 96tl Travel Photo International. 96bl GSF. 96tr Biofotos/H Angel. 96br Institute of Geological Science. 104 John Reader. 107 Spacecharts. 109t Michael Holford. 109b RHPL. 110, 111tl, 111tr Hutchison Library. 111bl NHPA/P Johnson. 111br RHPL. 112 PEP. 113l, 113c Hulton Picture Company. 113r Mary Evans Picture Library,. 114l Jacana. 114r Premaphotos/K Preston-Mafham. 114-115 Frank Lane Agency/Geoff Moon. 115 Premaphotos/K Preston-Mafham. 117t PEP/P Scoones. 117b Dr Jens Lorenz Franzen. 119 Dr LM Cook. 122 SPL/Eric Grave. 123l SPL/Dr T Brain & D Parker. 123r SPL Division of

Computer Research & Technology, National Institute of Health. 123r John Watney. 124 OSF/Peter Parks. 125 OSF/Barne E Watts. 126 Premaphotos/K Preston-Mafham. 127t GR Roberts. 127b NHPA/John Shaw. 128 Eric Crighton. 129l BCL/Hans Reinhard. 129r Iriate. 130 Biofotos/Heather Angel. 131 Graham Bateman. 133 BCL/Kim Taylor. 134 OSF/Peter Parks. 136t PEP/Nancy Sefton. 136b PEP/K Vaughan. 137l RW Van Devender. 137r A/P Morris. 140 Natural Science Photos/I Bennett. 141 Jacana/Chaumeton. 142 Tropical Australia Graphics. 143 Anthony Bannister. 144 Agence Nature/Chaumeton/Labat. 145 A/V Taylor. 150 A. 151 A/JP Ferrero. 155 CA Henley. 160 Telegraph Colour Library/Alex Lowe. 161 Ronald Sheridan/Ancient Art & Architecture Collection. 162t Zefa. 162b BCL/LC Marigo. 163 Zefa. 164 Holt Studios. 165 Art Directors/Chuck O'Rear. 166 ANT/Gordo Claridge. 169l NHP/A/N A Callow. 169r John & Penny Hubley. 169b BCL/Nicholas De Vore. 172l, 172r HL. 173t HL/John Downman. 173b Premaphotos Wildlife/KG Preston-Mafham. 174 HL/Dave Brincombe. 175l HL/Sarah Errington Pollock. 175r HL/Timothy Beddow. 176 HL/S Porlock. 177 BCL/LC Marigo. 163 Zefa. 164 Holt Studios. 165 Art Directors/Chuck O'Rear. 166 ANT/Gordon Claridge. 169l NHP/A/NACallow. 169r John & Penny Hubley. 169b BCL/Nicholas De Vore. 172l, 172r HL. 173t HL/John Bownman. 173b Premaphotos Wildlife/KG Preston-Mafham. 174 HL/Dave Brincombe. 175l HL/Sarah Errington Pollock. 175r HL/Timothy Beddow. 176 HL/S Porlock. 177 RHPL/GS Corrigan. 181 Aquila/ M & A Lane. 182 HL. 185 Ardea/JP Ferrero. 185 (inset) Swift Picture Library/R Fletcher. 186l A Mowlem. 186r BCL. 188t BCL/Cameron Davidson. 188b RHPL. 189t BCL/BD Hamilton. 189b Agence Nature/ JP Ferro. 190 PEP/J Duncan. 191 NHPA/Peter Johnson. 193, 194 RHPL. 196l PEP/J & G Lythgoe. 196r BCL/D Goulston. 197 BCL/K Gunnar. 198 NHPA/David Woodfall. 200t Premaphotos Wildlife. 200b PEP/J Scott. 201 Anthony Bannister. 204c NHPA/Harold Palo Jr. 204b Michael Fogden. 205tr Biofotos/Soames Summerhays. 205cl OSF/D Thompson. 205br NHPA/Harold Palo Jr. 206 PEP/Peter David. 207 PEP/Herwath Viogtmann. 208 Hutchison Library/Richard House. 209 Hutchison Library/Patricio Goycolea. 210 Zefa/B Crader. 212 SPL/Earth Satellite Corporation. 212-213 ANT 213 RHPL/John G Ross. 214 BCL/David Davies. 215t A Charnock. 215b Peter Fraenkel. 216t NHPA/David Woodfall. 216b Zefa. 216-217b Doug Weschler. 218 OSF/Rob Cousins. 219r OSF. 219l Aspect/Tom Nebbia. 222-223 BCL/Francisco Erize. 223l RHPL. 223r Peter Veit. 224t Biofotos/Heather Angel. 22b NHPA/Douglas Dickins. 225t George Frame. 225bl PEP/Arup & Marioj shah. 225br Ardea. 228 Survival Anglia/Jeff Foott. 229l Zefa. 229r, 230 Rex Features. 230-231 South American Pictures/Tony Morrison. 231l Biofotos/Heather Angel. 232 Colorific/Bill Pierce. 233l Novosti Press Agency. 233r Frank Spooner/Gamma. 234t, 235 Susan Griggs Agency/Martin Rodgers. 235b Zefa.